Annals of Mathematics Studies

Number 71

NORMAL TWO-DIMENSIONAL SINGULARITIES

BY

HENRY B. LAUFER

PRINCETON UNIVERSITY PRESS

AND

UNIVERSITY OF TOKYO PRESS

PRINCETON, NEW JERSEY

1971

LC Card: 78-160261
ISBN: 0-691-08100-x
AMS 1970: 32C40

Published in Japan exclusively by
University of Tokyo Press;
in other parts of the world by
Princeton University Press

Printed in the United States of America

To My Parents

PREFACE

This monograph is an outgrowth of a course given in 1969-70 at Princeton University. Its aim is to analytically describe and classify normal 2-dimensional singularities of complex spaces. By restricting considerations solely to dimension two, it is possible, in certain theorems, to get more detailed results than are known in the general case.

The reader should have a good knowledge of several complex variables and some acquaintance with Riemann surfaces and covering spaces.

I would like to thank the students who attended my course for their many helpful corrections and suggestions. I would also like to thank the secretaries at Fine Hall, especially Florence Armstrong and Elizabeth Epstein, for their aid in preparing the manuscript.

INTRODUCTION

The detailed study of normal 2-dimensional singularities is much easier than the higher dimensional case primarily for two reasons. First, any normal 2-dimensional singularity p is isolated. Thus, in resolving p, §II, [Hz1], we replace p by a compact analytic space A. Secondly, because p is 2-dimensional, A is 1-dimensional. The theory of compact Riemann surfaces gives a great deal of information about A and small neighborhoods of A. Since p is normal, p is determined by any small neighborhood U of A. It is U which we actually study.

Let $A = \cup A_i$ be the decomposition of A into irreducible components. Thus each A_i is a (possibly singular) Riemann surface. It is easy to reduce all considerations to the case where the A_i are non-singular, intersect transversely, and no three A_i meet at a point. Allowing regular points to also be "resolved", A comes from a resolution if and only if the intersection matrix of the A_i in U is negative definite, §IV, [M], [Gr2]. Moreover, it is easy to decide if two different A's can possibly resolve the same singularity, §V, [Ho], [B].

The next problem then is to get information about singularities with the same A and the same intersection matrix. This is done in §VI, [Gr2], [H & R], where it becomes necessary to introduce infinitesmal neighborhoods of A, or more precisely, analytic spaces with nilpotents having A as their underlying topological space. It is first shown that if A and $\tilde{A}$ have suitable isomorphic infinitesmal neighborhoods, then A and $\tilde{A}$ have formally equivalent neighborhoods. It is then shown that formal equivalence implies actual equivalence. Using Riemann-Roch, it becomes possible to get an estimate, in terms of the genera of the A_i and the intersection matrix, on which infinitesimal neighborhoods must be isomorphic in order for A and $\tilde{A}$ to have isomorphic neighborhoods.

Finally in §VII, [H & R], we obtain a complete set of invariants for p, namely the C-algebra structure of $\mathcal{O}_p/m^\lambda$, where m is the maximal ideal in $\mathcal{O}_p$ and λ is sufficiently large. Again, because of the dimension two condition, we can improve upon previously known results and get an estimate on λ in terms of the genera of the A_i and the intersection matrix. Thus the same degree of truncation suffices to determine all normal 2-dimensional singularities which have homeomorphic resolutions.

CONTENTS

Normal Two-Dimensional Singularities

CHAPTER I

RESOLUTION OF PLANE CURVE SINGULARITIES

Our primary tool for the study of singularities will be resolutions. Roughly speaking, in resolving singularities, we add more holomorphic functions and, if necessary, replace the singular points by larger sets in order to get a manifold. In this section we shall resolve the singularities of plane curves, i.e. hypersurfaces in 2-dimensional manifolds, via a canonical process.

DEFINITION 1.1. If V is an analytic space, a resolution of the singularities of V consists of a manifold M and a proper analytic map $\pi: M \to V$ such that π is biholomorphic on the inverse image of R, the regular points of V, and such that $\pi^{-1}(R)$ is dense in M.

DEFINITION 1.2. A quadratic transformation at a point p in a 2-dimensional manifold M consists of a new manifold M' and a map $\pi: M' \to M$ such that π is biholomorphic on $\pi^{-1}(M-p)$ and π is given near $\pi^{-1}(p)$ as follows.

Let (x,y) be a coordinate system for a polydisc neighborhood $\Delta(0;r) = \Delta$ of p, with $p = (0,0)$. $\Delta' = \pi^{-1}(\Delta)$ has two coordinate patches $U_1 = (u,v)$ and $U_2 = (u',v')$ with $u' = \frac{1}{u}$ and $v' = uv$. $U_1 \cap U_2 = \{u \neq 0\}$. $\pi(u,v) = (uv,v)$ and $\pi(u',v') = (v',u'v')$. Thus $\Delta = \{(x,y) \mid |x| < r_1, |y| < r_2\}$, $U_1 = \{(u,v) \mid |uv| < r_1, |v| < r_2\}$ and $U_2 = \{(u',v') \mid |v'| < r_1, |u'v'| < r_2\}$.

A quadratic transformation as defined above is often called a monoidal transformation, a σ-process, or a blowing-up.

π may be thought of as replacing p by the complex lines on M going through p. To see this, look at the image of a fibre $u = u_0$. (u_0v,v) is just a complex line segment through the origin. With this interpretation, a typical neighborhood N in M' of a point $q \in \pi^{-1}(p)$ consists of line segments on M near q in direction (which gives $N \cap \pi^{-1}(p)$) and those points on M-p which lie on $N \cap \pi^{-1}(p)$ (which gives $N \cap \pi^{-1}(M\text{-}p)$).

Quadratic transformations are canonical. Namely, let $\phi: M \to \tilde{M}$ be a biholomorphic map between the 2-dimensional manifolds M and $\tilde{M}$ and let $\tilde{\pi}: \tilde{M}' \to \tilde{M}$ be a quadratic transformation at $\phi(p)$. Then there is a unique induced biholomorphic map $\phi': M' \to \tilde{M}'$ such that $\phi \circ \pi = \tilde{\pi} \circ \phi'$.

Having obtained M' by a quadratic transformation, one can repeat the process and form M'' by a quadratic transformation at a point of M'.

Intuitively, one might also think of a quadratic transformation as spreading out or separating curves through the origin. Thus, for example, the curves $C_1 = \{x = 0\}$ and $C_2 = \{y = 0\}$ meet in $\mathbb{C}^2$, but $\overline{\pi^{-1}(C_1\text{-}0)} = \{u = 0\}$ and $\overline{\pi^{-1}(C_2\text{-}0)} = \{u' = 0\}$ do not meet. Thus the singularity of $V = \{xy = 0\}$ is resolved by $\pi: \overline{\pi^{-1}(C_1\text{-}0)} \cup \overline{\pi^{-1}(C_2\text{-}0)} \to V$. Similarly, look at $V = \{x^2 - y^3 = 0\}$. $\pi^{-1}(V) = \{u^2v^2 - v^3 = v^2(u^2 - v) = 0\} \cup \{v'^2 - u'^3v'^3 = v'^2(1 - u'v') = 0\}$. Of course $\pi^{-1}(0) \subset \pi^{-1}(V)$, but we may discard it and observe that $\pi: \overline{\pi^{-1}(V\text{-}0)} \to V$ is a resolution. Theorem 1.1 below says that repeated quadratic transformations as above will always resolve the singularities of plane curves.

THEOREM 1.1. *Let* V *be a 1-dimensional subvariety in a 2-dimensional manifold* M. *There exists a manifold* $\tilde{M}$ *obtained from* M *by successive quadratic transformations,* $\pi: \tilde{M} \to M$, *such that if* R *is the set of regular points on* V, $\pi: \overline{\pi^{-1}(R)} \to V$ *is a resolution of the singularities of* V. *Locally,* $\tilde{M}$ *is obtained from* M *by only a finite number of quadratic transformations.*

Proof: The singular points of V are isolated so the theorem is purely local in nature. We only have to resolve each singular point.

First suppose that the origin 0 is a singularity of the irreducible subvariety V in $\mathbb{C}^2$. By the local representation theorem (III. A.10 of G & R), we can choose coordinates near 0 so that $\rho: (x,y) \to x$ expresses

$V - \{0\}$ (in some neighborhood of 0) as a connected (since V is irreducible) s-sheeted covering space of a punctured disc $N - \{0\}$ in the x-plane. Since all s-sheeted coverings of $N - \{0\}$ are analytically the same, $\rho: V - \{0\} \to N - \{0\}$ is equivalent to $\tilde{\rho}: U - \{0\} \to N - \{0\}$ via $x = t^s$, where U is a disc in the t-plane. y is an analytic function of t on $U - \{0\}$. Since the equivalence of ρ and $\tilde{\rho}$ may be extended by mapping the origin to the origin, the Riemann removable singularity theorem insures that y is analytic in U. Hence $V = \{(x,y) \mid f(x,y) = 0\}$ can be represented as the image of U under $t \to (t^s, y(t))$. y(t) has a power series expansion $y(t) = a_m t^m + a_{m+1} t^{m+1} + \ldots$, $a_m \neq 0$. $t \to (t^s, y(t))$ is a one-to-one map.

No longer trying to preserve the covering map ρ, we now wish to find coordinates expressing V in a more useful manner as locally the image of a disc.

Multiplying y by $1/a_m$ and if necessary choosing a new parameter t and interchanging x and y, we may assume that V is locally given by the image of $t \to (t^s, t^m + a_{m+1}t^{m+1} + \ldots)$ with $m \geq s$.

s,m and those j such that $a_j \neq 0$ are relatively prime, for if $r \neq 1$ divides s,m and all the j's, $t \to (t^s, y(t))$ cannot be one-to-one since it factors through $t \to T = t^r \to (T^{s/r}, T^{m/r} + \ldots)$. We may assume that m is not a multiple of s, for if $m = ks$ let $(x',y') = (x, y-x^k)$. If $s = 1$, there is no singularity.

What happens under a quadratic transformation?

$$V = \{(x,y) \mid f(x,y) = a_{10}x + a_{01}y + a_{20}x^2 + \ldots = 0\}.$$

$$\pi^{-1}(V) = \{(u,v) \mid f^*(u,v) = a_{10}uv + a_{01}v + a_{20}u^2v^2 + \ldots = 0\}$$

$$\cup \{(u',v') \mid a_{10}v' + a_{01}u'v' + a_{20}u'^2 + \ldots = f^*(u',v') = 0\}.$$

If k is the order of the zero of f(x,y) at (0,0), we may factor v^k out of $f^*(u,v)$ and $f^*(u',v')$. Hence f^*/v^k defines the subvariety $V' = \overline{\pi^{-1}(V - \{0\})}$.

On $M' - \pi^{-1}(\{x = 0\})$, $u' = y/x$ and $v' = x$. Hence $\pi^{-1}(V - \{0\})$ is given by $t \to (u',v') = (t^{m-s} + a_{m+1}t^{m-s+1} + \ldots, t^s)$ which extends to

$0 \to (0,0)$. Thus $u' = v' = 0$ is the only possible singularity of V'. If $m - s > s$, perform another quadratic transformation at $u' = v' = 0$. Eventually $m - \mu s < s$ and we can choose new parameters for $V^{(\mu)}$, $t \to (t^{s'}, t^{m'} + \ldots)$ with $s' < s = m'$.

We can now perform more quadratic transformations. Eventually $s^{(\nu)} = 1$ and we have a non-singular $V^{(\nu)}$.

In the general case, the singularity of V at 0 will have a finite number of irreducible components. Perform quadratic transformations until these components are all desingularized. We now have to separate the components. As observed before, if two components intersect transversely (i.e. have distinct tangent planes), a single quadratic transformation at their point of intersection separates them. Suppose V_1 and V_2 are non-singular and tangent at $0 = V_1 \cap V_2$. We may suppose, after an appropriate choice of coordinates, that $V_1 = (t,0)$ and $V_2 = (t, t^m + a_{m+1}t^{m+1} + \ldots)$. The proof is by induction on m. If $m = 1$, V_1 and V_2 meet transversely. If $m > 1$, after a quadratic transformation, $V_1' = (t,0)$ and $V_2' = (t, t^{m-1} + a_{m+1}t^m + \ldots)$, which proves the induction step.∎

We observe for use later the following.

Proposition 1.2. *If V is a non-singular 1-dimensional subvariety near the origin of* $\mathbf{C}^2$ *and* $\pi : M' \to \mathbf{C}^2$ *is a quadratic transformation at the origin, then* $\overline{\pi^{-1}(V - \{0\})}$ *and* $\pi^{-1}(0)$ *intersect transversely.*

CHAPTER II

RESOLUTION OF SINGULARITIES OF TWO-DIMENSIONAL ANALYTIC SPACES

THEOREM 2.1. *Any 2-dimensional analytic space has a resolution.*

Theorem 2.1 is really a local theorem, but we do not know that yet. Anyway, we begin by proving a local version.

PROPOSITION 2.2. *Let* $p \in S$, *a 2-dimensional analytic space. Then there is a resolution* $\pi: M \to V$ *of some neighborhood* V *of* p.

Proof: We may assume that p is a singular point. S has a finite number $V_1, \ldots, V_s$ of irreducible components near p. It will suffice to resolve each V_i separately. We may assume that $V = V_i$ is locally given as an admissible presentation as in III. A.10 of G & R, i.e.

$$\rho: V \to \{z_1, z_2\}$$

represents V as an analytic cover. Let $B = \text{loc } D$ denote $\rho(V(D))$, i.e. the image of the locus of the discriminant, which contains all points above which ρ may fail to be a covering map. B is the branch locus if ρ is thought of as a branched covering map. B is a plane curve. Choose a small enough neighborhood N of the origin so that $0 = (0,0)$ is the only possible singularity of B. By Theorem 1.1, with a finite number of quadratic transformations in the (z_1,z_2)-plane, we can resolve the singularity of B. Quadratic transformations do not depend on the coordinate system, so we may assume that they are given by mappings of the form $\pi: N' \to N$, $\pi: (u,v) \to (uv,v) = (z_1,z_2)$ and $\pi: (u',v') \to (v',u'v') =$

$(z_1 z_2)$. If V is locally embedded in C^r, π may be extended to

$$\tilde{\pi}: (u,v,z_3,\ldots,z_r) \to (uv,v,z_3,\ldots,z_r)$$

$$\tilde{\pi}: (u',v',z_3,\ldots,z_r) \to (v',u'v',z_3,\ldots,z_r)$$

Let $V' = \tilde{\pi}^{-1}(V)$. V' is a subvariety of $\pi^{-1}(C^r)$ and we have the following commutative diagram:

(2.1)

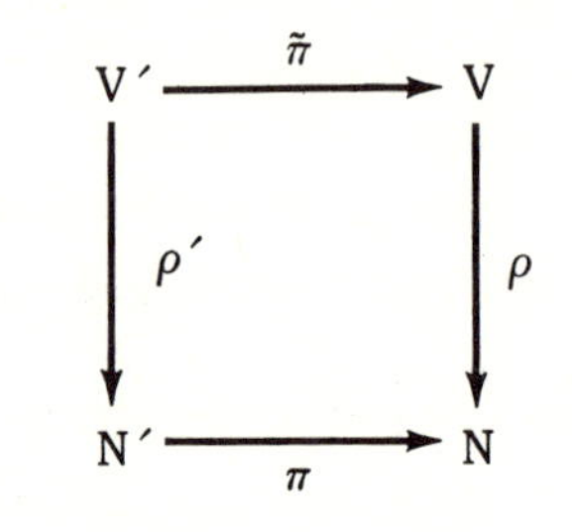

$\tilde{\pi}$ is biholomorphic except for $\tilde{\pi}^{-1}(\rho^{-1}(0)) = \tilde{\pi}^{-1}(0)$, a projective line. Thus $\rho': V' \to N'$ represents V' as an analytic cover with $\pi^{-1}(B) = B'$ as branch locus.

Thus when we resolve the singularities of B, we get successive diagrams of the form of (2.1). We induce a change in V and get a new analytic cover with $B' = \pi^{-1}(B) = \pi^{-1}(0) \cup \overline{\pi^{-1}(B-\{0\})}$ ($\overline{\pi^{-1}(B-\{0\})}$ is now a submanifold) as branch locus. We want to be able to choose local coordinates (ζ_1,ζ_2) such that the singularities of $\pi^{-1}(B)$ are only of the form $\{\zeta_1\zeta_2 = 0\}$. Hence more quadratic transformations are required. First we want the irreducible components $\{B_j'\}$ of B' to intersect transversely. Since all of the B_j' are non-singular, by Proposition 1.2, if we perform a quadratic transformation ω at a point $q \in B_j'$, $\overline{\omega^{-1}(B_j' - q)}$ (also denoted by B_j') and $\omega^{-1}(q)$ meet transversely. If B_j' and B_k', $j \neq k$, do not meet transversely at a point q, apply Theorem 1.1 to resolve the singularity at q. B_j' and B_k' will then not meet at all.

We now have all the irreducible components of B meeting transversely. It may still happen, however, that three or more B_i meet a single point q. A quadratic transformation at q separates the B_i', which will now meet $\pi^{-1}(q)$ transversely at distinct points.

So far, by quadratic transformations we have brought V' to the point where it is represented as an analytic cover in such a manner that locally the branch locus is either a submanifold or can be given by $\{\zeta_1\zeta_2 = 0\}$ in some polydisc Δ. We shall resolve the latter case first. $(\rho')^{-1}(\Delta - \{\zeta_1\zeta_2 = 0\})$ is a finite sheeted covering space of $\Delta - \{\zeta_1\zeta_2 = 0\}$ and each connected component has an irreducible subvariety as its closure (III. C. 20 of G & R). These irreducible components are also represented by ρ' as analytic covers with $\{\zeta_1\zeta_2 = 0\}$ as branch locus. We shall resolve these irreducible singularities using the manifolds $M(k_1, \ldots, k_s)$ below.

Given $k_1, \ldots, k_s$ with the k_i integers such that $k_i \geq 2$, $M = M(k_1, \ldots, k_s)$ will be covered by $s+1$ coordinate patches, $U_i = (u^{(i)}, v^{(i)}) = \mathbf{C}^2$, $0 \leq i \leq s$, joined as follows.

$$U_0 \cap U_1 = \{u \neq 0\} \qquad u' = \frac{1}{u} \qquad v' = u^{k_1}v$$

$$U_1 \cap U_2 = \{v' \neq 0\} \qquad v'' = \frac{1}{v'}, \qquad u'' = v'^{k_2}u'$$

$$U_2 \cap U_3 = \{u'' \neq 0\} \qquad u''' = \frac{1}{u''} \qquad v''' = (u'')^{k_3}v''$$

$$\vdots$$

$$U_{2n} \cap U_{2n=1} = \{u^{(2n)} \neq 0\} \qquad u^{(2n+1)} = \frac{1}{u^{(2n)}} \qquad v^{(2n+1)} = (u^{(2n)})^{k_{2n+1}}v^{(2n)}$$

$$U_{2n+1} \cap U_{2n+2} = \{v^{(2n+1)} \neq 0\} \quad v^{(2n+2)} = \frac{1}{v^{(2n+1)}} \quad u^{(2n+2)} = (v^{2n+1)})^{k_{2n+2}}u^{(2n+1)}$$

$$\vdots$$

$A = \{v = v' = 0\} \cup \{u' = u'' = 0\} \cup \{v'' = v''' = 0\} \cup \ldots$ is a compact analytic subset of M. $(U_0 \cap U_1) - A = \{uv \neq 0\}$ undergoes an automorphism at each change of coordinates. Thus $M = \{uv \neq 0\} \cup A \cup \{v^{(s)} = 0\}$, if s is even and $M = \{uv \neq 0\} \cup A \cup \{u = 0\} \cup \{u^{(s)} = 0\}$, if s is odd. Let us suppose that s is odd. If s is even, $v^{(s)}$ must be replaced by $u^{(s)}$ in the following arguments.

v and $v^{(s)}$ are in fact analytic functions on all of M. To see this, we must see that they are analytic in each coordinate patch.

$$v^{(s)} = (u^{(s-1)})^{k_s} v^{(s-1)} = (u^{(s-2)})^{k_s} (v^{(s-2)})^{k_{s-1}k_s-1}$$

$$= (u^{(s-3)})^{k_{s-2}(k_{s-1}k_s-1)-k_s} (v^{(s-3)})^{k_{s-1}k_s-1}$$

$$= (u^{(s-4)})^{k_{s-2}(k_{s-1}k_s-1)-k_s} (v^{(s-3)})^{k_{s-3}[k_{s-2}(k_{s-1}k_s-1)-k_s]-(k_{s-1}k_s-1)}$$

$$= \ldots$$

We must check that the exponents are all positive. In each coordinate patch, alternately dividing the first exponent by the second exponent and then dividing the second by the first, we get successively,

$$k_s,\ k_{s-1} - \frac{1}{k_s},\ \ k_{s-2} - \frac{1}{k_{s-1} - \frac{1}{k_s}},\ \ k_{s-3} - \frac{1}{k_{s-2} - \frac{1}{k_{s-1} - \frac{1}{k_s}}},$$

$\ldots,$

i.e. continued fractions, since $k_i \geq 2$. An easy induction proof shows that these ratios are all greater than 1. Hence $v^{(s)}$, and similarly v, is analytic on M. Moreover, $v^{(s)} = u^a v^b$ with $a > b$, a and b relatively prime and

$$\frac{a}{b} = k_1 - \frac{1}{k_2 - \frac{1}{k_3}} \quad \ldots\ .$$

If s is even, $u^{(s)} = u^a v^b$ with, as before, $a > b$, a and b relatively prime and

$$\frac{a}{b} = k_1 - \frac{1}{k_2 - \frac{1}{k_3}} \quad \ldots\ .$$

If σ and ω are positive integers,

$$\rho : \{uv \neq 0\} \to (v^{\sigma}, (u^a v^b)^{\omega}) = (\zeta_1, \zeta_2) \tag{2.2}$$

is a covering map onto $\zeta_1\zeta_2 \neq 0$ which may be extended to make M an analytic cover of C^2 with $\{\zeta_1\zeta_2 = 0\}$ as branch locus. This will be exactly what we need in order to resolve the singularity currently under consideration.

Thus, returning to the main line of our argument, we have a part W of V′ represented by ρ' as an irreducible analytic cover in a polydisc neighborhood Δ of the origin with $\zeta_1\zeta_2 = 0$ as the branch locus. We first classify ρ' on $(\rho')^{-1}(\zeta_1\zeta_2 \neq 0)$, where ρ' is a covering map. The injection of $\Delta - \{\zeta_1\zeta_2 \neq 0\}$ into $C^2 - \{\zeta_1\zeta_2 = 0\}$ is a homotopy equivalence. Thus analytic covering spaces of $\Delta - \{\zeta_1\zeta_2 = 0\}$ extend naturally to analytic covering spaces of $C^2 - \{\zeta_1\zeta_2 = 0\}$ and it suffices to classify the latter and then restrict back to $\Delta - \{\zeta_1\zeta_2 = 0\}$. Covering spaces correspond to subgroups of the fundamental group. $C^2 - \{\zeta_1\zeta_2 = 0\}$ may be continuously deformed onto $|\zeta_1| = |\zeta_2| = 1$, a torus. Hence $\pi_1 = \pi_1(C^2 - \{\zeta_1\zeta_2 = 0\}) = Z \oplus Z$. W is a finite sheeted covering space and thus corresponds to a subgroup $G \subset \pi_1$ of finite index. Thus $G \approx Z \oplus Z$ is determined by its two generators in π_1. Thus ρ' is analytically equivalent to a covering ρ' of the form

$$\rho' : C^2 - \{\xi\eta = 0\} \to C^2 - \{\zeta_1\zeta_2 = 0\} \tag{2.3}$$

with $\rho'(\xi,\eta) = (\xi^{\alpha}\eta^{\beta}, \xi^{\gamma}\eta^{\delta})$ with $\alpha, \beta, \gamma, \delta$ integers such that $\alpha\delta - \beta\gamma \neq 0$.

Automorphisms of the covering space $C^2 - \{\xi\eta = 0\}$ will not affect the validity of the cover and will enable us to put ρ' into the form of (2.2). $(\xi,\eta) \to (\xi', \eta') = (\xi, \xi^{\nu}\eta)$ and $(\xi,\eta) \to (\xi'', \eta'') = (\xi\eta^{\mu}, \eta)$ with ν and μ integers represent ρ' as $\rho'(\xi',\eta') = (\xi'^{\alpha}\eta'^{\beta-\nu\alpha}, \xi'^{\gamma}\eta'^{\delta-\nu\gamma})$ and $\rho'(\xi'',\eta'') = (\xi''^{\alpha-\mu\beta}\eta''^{\beta}, \xi''^{\gamma-\mu\delta}\eta''^{\delta})$ respectively. We also have the automorphisms $(\xi,\eta) \to (\frac{1}{\xi}, \eta)$ and $(\xi,\eta) \to (\eta,\xi)$ at our disposal.

Using the Euclidean algorithm on α and β, as induced by appropriate automorphisms, we may put (2.3) into the form (omitting primes)

$$\rho': (\xi, \eta) \to (\xi^\sigma, \xi^\lambda \eta^\tau)$$

where σ is the greatest common divisor of $|\alpha|$ and $|\beta|$, and $\tau > \lambda \geq 0$. If ω is the greatest common divisor of τ and λ, we have

$$\rho': (\xi, \eta) \to (\xi^\sigma, (\xi^a \eta^b)^\omega)$$

with a and b relatively prime and $b > a \geq 0$. Any $\frac{b}{a}$, $a \neq 0$, can be written (uniquely) as a finite continued fraction of the form

$$\frac{b}{a} = k_1 - \cfrac{1}{k_2 - \cfrac{1}{k_3 \, \ddots \, - \cfrac{1}{k_s}}}$$

with $k_i \geq 2$. Now restrict ρ' to $\rho'^{-1}(\Delta - \{\zeta_1\zeta_2 = 0\})$. It is of the form (2.2). Thus, after the appropriate restriction of ρ, there is a biholomorphic map π making the following diagram commutative. $a = 0$ corresponds to $M(\emptyset) = \mathbf{C}^2$.

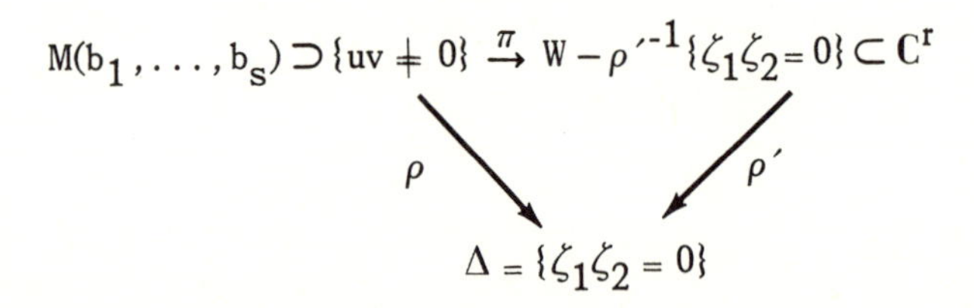

In local coordinates, π is locally bounded and hence by the Riemann removable singularity theorem extends to an analytic map $\pi: M \to W$ giving the following commutative diagram:

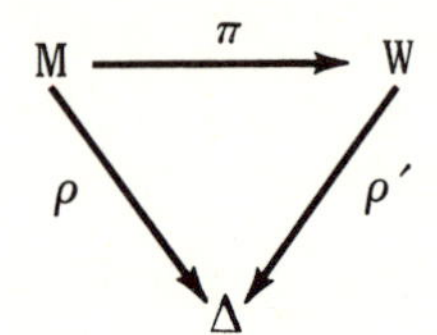

ρ is a proper map since $\{|v| \leq 1\}$ $\{|v^{(s)}| \leq 1\}$ is compact in M. Thus π is a proper map. Thus π is a resolution of W except possibly that π is not biholomorphic at manifold points of W which lie over the branch locus.

Let us first determine the nature of π near the pre-image of a point $q \in W$ such that $\rho'(q) = (q_1,q_2)$ with $q_1 \neq 0$ and $q_2 = 0$. $\rho^{-1}(\rho'(q))$ consists of σ distinct points, $q', q'', \ldots, q^{(\sigma)}$, on the v-axis in the U_0 coordinate patch. ρ represents a neighborhood $N^{(j)}$ of each of the $q^{(j)}$ as an analytic cover with $\zeta_2 = 0$ as branch locus. Also, ρ restricted to $N^{(j)}$ is one-to-one on the branch locus. Since π is biholomorphic off $\rho^{-1}(\zeta_1\zeta_2 = 0)$, $\pi(\cup N^{(j)})$ will have exactly σ connected components off $(\rho')^{-1}(\zeta_1\zeta_2 = 0)$. Hence W has exactly σ irreducible components above (q_1,q_2). Some of these will meet at q. If W′ is an irreducible component at q, there will be one N′ which corresponds to W′. $\pi : N' \to W'$ is biholomorphic off the branch locus. Since ρ is one-to-one above the branch locus, π is also one-to-one above the branch locus. Thus π is a homeomorphism.

Summarizing, the map π is obtained by breaking W up into irreducible components and then resolving each component by a homeomorphism. If q is a manifold point, there is only one irreducible component at q and π^{-1} is a homeomorphism which is holomorphic except possibly above the branch locus. But then the Riemann removable singularity theorem shows that π^{-1} is holomorphic.

We must show that our locally defined $\pi_i : M_i \to W_i$ agree for intersecting W_i on V′. But according to the previous paragraph, whenever the branch locus of $\rho' : V' \to N'$ is a submanifold (which is everywhere except for a discrete set), π_i is obtained by decomposing V′ into irreducible components and then resolving each component by a homeomorphism π_i. Given two different resolutions π_i and π_j, $\pi_i^{-1} \circ \pi_j$ is a homeomorphism and biholomorphic except on a proper subvariety. Thus $\pi_i^{-1} \circ \pi_j$ is biholomorphic on $W_i \cap W_j$. The π_i patch together to give a manifold X and $\pi : X \to V'$ such that π is proper, holomorphic and biholomorphic on the $x \in X$ such that $\rho'(\pi(x))$ is not in the branch locus. Also, except above the singular points of the branch locus, π is biholomorphic above the manifold points of V'.

Recall that $\tilde{\pi}: V' \to V$ is proper, holomorphic and biholomorphic off $\tilde{\pi}^{-1}(p)$. p has been assumed to be singular. Thus $\tilde{\pi} \circ \pi : X \to V$ is a resolution of the singularities of V. ▮

We can now prove Theorem 2.1.

Let S be the given analytic space. Consider the set of T of all points $t \in S$ such that for some neighborhood V of t, a resolution π may be chosen so that for all $q \in V$, π resolves each irreducible component of V_q by a homeomorphism. By the proof of Proposition 2.2, S-T is discrete. Moreover, the local resolutions near points in S-T will patch together with the resolutions near points in T to give a global resolution of S. ▮

If $\pi: M \to V$ is a resolution of a 2-dimensional analytic space constructed as in the proof of Theorem 2.1, then $\pi^{-1}(p)$ is 0-dimensional except for a discrete set of points p in V. If $\pi^{-1}(p)$ is 1-dimensional, then the irreducible components of $A = \pi^{-1}(p)$ are just (singular) compact Riemann surfaces. By Theorem 1.1, by performing quadratic transformations at points of A, and thereby getting a different resolution for V, we can always find resolutions such that A consists of non-singularly embedded Riemann surfaces which intersect transversely and such that no three intersect at a point. It is customary to represent A by its dual graph Γ as follows. Let $\{A_i\}$ be the irreducible components of A. These A_i are the vertices of the graph. An edge connecting two vertices A_i and A_j corresponds to a point of intersection of the Riemann surfaces A_i and A_j. Each Riemann surface A_i represents a topological homology class in M and thus has a well-defined self-intersection number $A_i \cdot A_i$, which may also be defined as the Chern class of the normal bundle of the embedding. See Theorem 2.3 below.

We may assume that M is oriented so that a transverse intersection of two submanifolds contributes $+1$ to their intersection number. To each vertex A_i of the graph Γ, we associate $A_i \cdot A_i$ and thus obtain a weighted graph, also denoted Γ. Thus the weighted graph associated to $M(k_1, \ldots, k_s)$ is

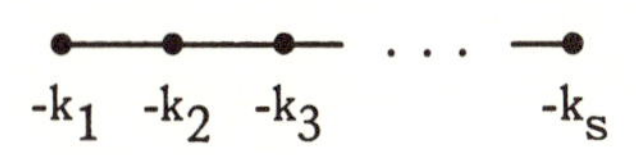

where each vertex is a projective line.

We need not use Theorem 2.3 subsequently. It would always suffice to define $A \cdot A$ as the Chern class of the normal bundle rather than as the topological self-intersection number.

THEOREM 2.3. *Let* N *be the normal bundle of a non-singular compact Reimann surface* A *embedded in the 2-dimensional manifold* M. *Then* $A \cdot A$ *equals the Chern class of* N.

Theorem 2.3 follows immediately from Lemmas 2.4 and 2.5 below.∎

LEMMA 2.4. *Let* D *be the* 0*-section of a line bundle* N *over a compact Riemann surface* A. *Then* c, *the Chern class of* N, *equals* $D \cdot D$.

Proof: If f is a non-trivial meromorphic section of N, which exists by [Gu, p. 107], then c equals the algebraic sum of the number of zeros and poles of f. If we had an analytic section $s: A \to N$, then s would be homologous to D and to compute $D \cdot D$, we could just count the number of intersections of s with D, i.e. the number of zeros of s. For f, look at a point $q \in A$ where f has a pole. We may choose local coordinates on N so that $q = (0,0)$, $N = \{(z_1, z_2) \in C^2 \mid |z_1| < 1\}$, $D = \{z_2 = 0\}$. f becomes a meromorphic function $z_2 = f(z_1)$ with a pole at z_1. Thus we may choose the z_1 coordinate so that $f(z_1) = \frac{1}{z_1^\nu}$, where ν is the order of the pole of f, $|z_1| \leq \varepsilon$. On $|z_1| = \varepsilon$, $f(z_1) = \frac{\bar{z}_1^\nu}{\varepsilon^{2\nu}}$. Thus we may get a piecewise differentiable section $\hat{f}$ on N by letting $\hat{f} = f$ for points where $|z_1| > \varepsilon$ and

$\hat{f} = \frac{\overline{z_1}^{\nu}}{\varepsilon^{2\nu}}$ for $|z_1| \leq \varepsilon$. Do this for all the pole points. $\hat{f}$ is homologous to D. Thus to compute $D \cdot D$, we can algebraically sum the number of zeros of $\hat{f}$. For a νth order zero of f, $f = z_1^{\nu}$, $|z_1| \leq \varepsilon$, we may approximate the analytic function z_1^{ν} on $|z_1| \leq \varepsilon$ by an analytic function $\tilde{f}$ having ν simple zeros. $\tilde{f}$ may be further modified, non-analytically, so as to agree with f for $|z_1| \geq \varepsilon$ and thus still be a section. Thus a νth order zero of f contributes ν to the number of zeros and poles. For a pole of f, $\hat{f} = \frac{\overline{z_1}^{\nu}}{\varepsilon^{2\nu}}$. As before, we approximate the anti-holomorphic function $\overline{z_1}^{\nu}$ by an anti-holomorphic function with ν simple zeros. A simple zero of an anti-holomorphic function contributes -1 to the intersection number of $\tilde{f}$ and D. Thus a νth order pole of f contributes $-\nu$ to the number of zeros of $\tilde{f}$. ∎

LEMMA 2.5. *Let* A *be a non-singularly embedded compact Riemann surface in the 2-dimensional manifold* M. *Let* D *be the* 0*-section of* N, *the normal bundle to* A. *Then* $A \cdot A = D \cdot D$.

Proof: Let $\hat{f}$ be the section of N constructed in Lemma 2.4. $\hat{f}$ meets D transversely at a discrete set of points.

Let us recall how N is defined. Let ${}_AT$ be the tangent bundle to A. Let ${}_MT$ be the restriction to A of the tangent bundle to M. $\iota: A \to M$ induces an injection of ${}_AT$ into ${}_MT$ and N is defined as the quotient bundle, i.e. (2.4) is exact.

$$0 \to {}_AT \xrightarrow{\iota} {}_MT \xrightarrow{\rho} N \to 0 \tag{2.4}$$

We wish to show that (2.4) splits as a sequence of differentiable bundles, i.e. there is a C^{∞} bundle map $h: N \to {}_MT$ such that $\rho \circ h$ is the identity. h will be complex linear on each fibre.

(2.4) yields (2.5), a new exact sequence of vector bundles. Exactness is just exactness on each fibre and Hom is exact for vector spaces.

(2.5) $$0 \to \mathrm{Hom}(N, {}_AT) \to \mathrm{Hom}(N, {}_VT) \to \mathrm{Hom}(N,N) \to 0$$

Using script letters, $\mathcal{H}om$, to denote the sheaf of germs of C^∞ sections of the vector bundle sequence (2.5), we get the following exact sheaf sequence:

$$0 \longrightarrow \mathcal{H}om\,(N, {}_AT) \xrightarrow{\iota *} \mathcal{H}om\,(N, {}_VT) \xrightarrow{\rho *} \mathcal{H}om\,(N,N) \longrightarrow 0$$

$H^1(A, \mathcal{H}om\,(N, {}_AT)) = 0$ since $\mathcal{H}om\,(N, {}_AT)$ is locally free and thus a fine sheaf. Hence there is a section $h \in \Gamma(A, \mathcal{H}om(N, {}_VT))$ such that $\rho_*(h)$ is the identity map. h is the desired splitting of (2.4). The image of h is a sub-bundle of ${}_MT$ whose fibres meet the tangent bundle of A transversely. Put another way, h chooses a transverse direction to A at each point.

Using h, we wish to transfer $\tilde{f}$ of Lemma 2.4 from N to M. Triangulate A finely enough so that each closed triangle T_i of the triangulation can be chosen to lie in a coordinate patch U_i of M with $U_i = \{(z_1,z_2) \in C^2 \,|\, |z_1| < 1, |z_2| < 1\}$ and $A \cap U_i = \{z_2 = 0\}$. Also, the zeros of $\tilde{f}$ should lie in the interior of the T_i. We can "graph" $\tilde{f}$, up to constant factor, on T_i as follows. At each point p of T_i, $h \circ \tilde{f}$ specifies a tangent vector. If we identify C^2 with the tangent space to C^2 at p, we get a point $h \circ \tilde{f}(p) \in C^2$. By multiplying $h \circ \tilde{f}(p)$ by a suitably small positive constant factor, k_i, we may assume that $k_i h \circ \tilde{f}(p) \in U_i$ for all $p \in T_i$. Let $G_i = k_i h \circ \tilde{f}(T_i) \subset M$. G_i meets A precisely at the zeros of $\tilde{f}$ and the intersections are transverse. G_i may be deformed continuously onto T_i. Unfortunately $\cup G_i$ is not necessarily a cycle since above $T_i \cap T_j$, G_i and G_j may differ. We correct this and form a cycle $\tilde{G}$ homologous to A with $\tilde{G} \cdot A = D \cdot D$ as follows. If $r \in \partial T_i$, the boundary of T_i, add to the points of G_i those points of U_i which lie on the real line segment joining $h \circ \tilde{f}(r)$ and r. Let $\tilde{G}_i$ be the union of G_i and all of these line segments for $r \in \partial T_i \cdot \tilde{G}_i$ is naturally the image of a triangle $\tilde{\sigma}_i$ with $\partial \tilde{\sigma}_i$ being mapping onto ∂T_i. Just take the triangle σ_i mapping onto T_i, map σ_i to G_i by $h \circ \tilde{f}$ and then add the line segments to the outside of σ_i, as illustrated below, to get $\tilde{\sigma}_i$.

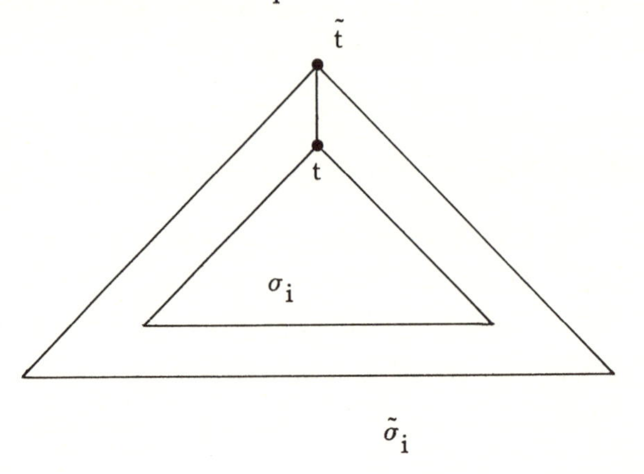

The line segment connecting t and $\tilde{t}$ is mapped linearly to the line segment connecting $h \circ \tilde{f}(t) \in M$ and $t \in M$.

$\tilde{G} = \cup \tilde{G}_i$ is then a homology cycle in M which is homologous to A. In the interiors of the T_i, $\tilde{G}$ and A meet only at the zeros of $\tilde{f}$ and these are transversal intersection points whose algebraic sum is $D \cdot D$. In addition, A and $\tilde{G}$ meet along the edges of the triangulation. It thus suffices to show that the intersection at ∂T_i contributes nothing to $\tilde{G} \cdot A$.

Near a point $r \in \partial T_i$, in the U_i coordinate system, $\tilde{G}$ has, in addition to the real line segment $L_i(r)$ connecting r and $h \circ \tilde{f}(r)$, only those $L_j(r)$ which come from a U_j coordinate system containing a triangle T_j such that $r \in T_j$. There are at most two other such T_j. $L_j(r)$ will be a curve in the U_i coordinate system, but will be tangent to $L_i(r)$ at r since its tangent at r is just $h \circ \tilde{f}(r)$. Thus if we move the points p near ∂T_i a small distance in the direction of $\tilde{f}(p)$, we will eliminate the intersection of A with $\tilde{G}$ near ∂T_i. In a U_j coordinate system, the points near T_i will have been moved a small distance in a direction close to $\tilde{f}(p)$ and thus away from $\tilde{G}$. Thus the additional movement needed in the U_j coordinate system will continue to move A away from $\tilde{G}$. Thus no new intersections are created and the lemma is proved. ∎

Suppose that A is a non-singular analytic submanifold of codimension 1 in the complex manifold M. Let $\mathcal{I}$ be the ideal sheaf of germs of holomorphic functions which vanish on A. Then $\mathcal{I}/\mathcal{I}^2$ is a locally free sheaf of rank 1 over A. (This is easily checked using local coordinate systems.) If N is the normal bundle of the embedding of A, then the line bundle of which $\mathcal{I}/\mathcal{I}^2$ is the sheaf of germs of sections is in fact N^*. To see this, we need to exhibit a canonical pairing between the fibres of $\mathfrak{N}$ and $\mathcal{I}/\mathcal{I}^2$ over a point $x \in A$. $<\mathfrak{N}_x,(\mathcal{I}/\mathcal{I}^2)_x> \to C$ may be given as follows. Using (2.4), if $n \in \mathfrak{N}_x$, then $n = \rho(t)$, locally, where $t \in {}_M\mathcal{T}_x$. If $g \in \mathcal{I}_x$, $<t,g>$ may be found by evaluating the tangent vector defined by t at x on the function g. We have to show that this pairing is independent of the choice of t and g. Any other choice of t will differ from t by a $t' \in {}_A\mathcal{T}_x$. But $g \equiv 0$ on A. Hence $<t',g> = 0$. Any other choice of g will differ from g by a function $g' \in \mathcal{I}_x^2$. Then $g' = \Sigma f_1 f_2$ with $f_1, f_2 \in \mathcal{I}_x$. But $<t, f_1 f_2> = 0$ since derivations vanish on products of functions which vanish at x. Hence $\mathfrak{N}^* \approx \mathcal{I}/\mathcal{I}^2$. Taking tensor products,

$$\mathcal{I}^m/\mathcal{I}^{m+1} \approx \mathcal{I}/\mathcal{I}^2 \otimes \ldots \otimes \mathcal{I}/\mathcal{I}^2 \approx \mathfrak{N}^* \otimes \ldots \otimes \mathfrak{N}^* \approx \bigotimes_m \mathfrak{N}^* \approx (\bigotimes_m \mathfrak{N})^*.$$

Suppose that f is a meromorphic function on a complex 2-dimensional manifold M. Let D be the zero and pole set of f. Let $\{D_i\}$ be the global irreducible components of D and let k_i be the order of the zero of f on D_i. We shall write $(f) = \Sigma k_i D_i$ and call this the divisor of f. Suppose that A is a compact, non-singular, 1-dimensional analytic subset of M and that $A = \cup A_j$ is its decomposition into connected components. Suppose also that the D_i are non-singular near A and D_i and A_j, all i and j, either meet transversely or coincide. In this case, we define the intersection number $A \cdot (f)$ as $\Sigma_j(\Sigma_i k_i A_j \cdot D_i)$ where $A_j \cdot D_i$ equals the number of intersection points of A_j with D_i if $A_j \neq D_i$ and $A_j \cdot D_i = A_j \cdot A_j$ if $A_j = D_i$.

THEOREM 2.6. *Let* (f) *be the divisor of the meromorphic function* f *on the* 2-*dimensional complex manifold* M. *Let* A *be a compact non-singular* 1-*dimensional analytic subset of* M. *Then if* $A \cdot (f)$ *may be defined as above,* $A \cdot (f) = 0$.

Proof: We may assume that A is connected. Let k be the order of the zero of f on A. Replacing f by 1/f if necessary, we may assume that $k \geq 0$. Let $\mathcal{J}$ be the ideal sheaf of germs of functions vanishing on A. Then f defines a section $\Gamma(A,\mathcal{J}^k)$ and hence a section $\tilde{f} \in \Gamma(A,\mathcal{J}^k/\mathcal{J}^{k+1}) \approx \Gamma(A, \otimes_k \mathcal{N}^*)$. By [Gu, p. 103], the algebraic sum of the zeros and poles of $\tilde{f}$ is $c(\otimes_k N^*) = -kc(N)$. But $\tilde{f}$ has a zero or pole precisely where A meets some D_i, $D_i \neq A$, and the order of the zero is precisely k_i. Hence $-kc(N) = \Sigma'_i \, k_i A \cdot D_i$ where Σ' indicates that we do not include $D_i = A$ in the summation. Then

$$(f) \cdot A = \Sigma_i \, k_i A \cdot D_i = \Sigma'_i \, k_i A \cdot D_i + kA \cdot A = -kc(N) + kA \cdot A =$$

$$= -kc(N) + kc(N) = 0 \;. \;\blacksquare$$

We now have sufficient machinery to calculate some examples. We compute our examples by explicitly following the steps described in the proof of Theorem 2.1.

Under $\rho: (x,y,z) \to (x,y)$, $z^2 = xy$ has branch locus $xy = 0$, as needed. We must determine the nature of the covering space over $xy \neq 0$. Restricting to $|x| = |y| = 1$, the universal covering space in the (θ,ϕ) plane with covering map $x = e^{2\pi i \theta}$, $y = e^{2\pi i \phi}$. The universal covering map then factors through the covering map $\rho: \{z^2 = xy\} \to (x,y)$ by $(x,y,z) = (e^{2\pi i\theta}, e^{2\pi i\phi}, e^{\pi i\theta + \pi i\phi})$. The inverse image of (1,1,1) is a lattice which determines the deck transformations and hence the covering map from the (θ,ϕ) plane to $z^2 = xy$, over $|x| = |y| = 1$.

The lattice is pictured below.

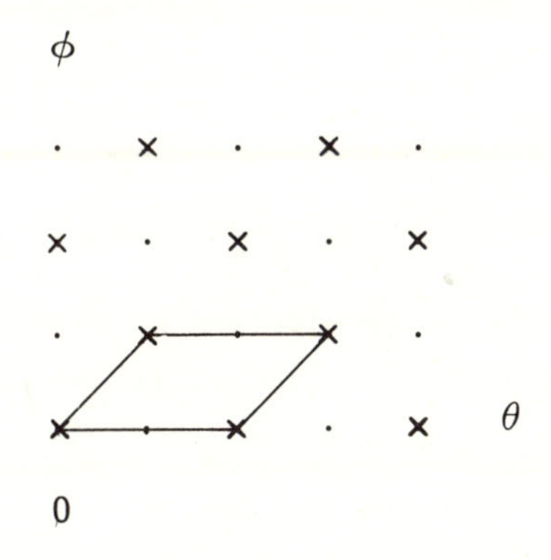

Choosing a fundamental parallelogram with vertices (0,0), (2,0), (1,1) and (2,1), we see that an equivalent covering is $(u,v) = (e^{\pi i\theta - \pi i\phi}, e^{2\pi i\phi})$. The map from (u,v) to $z^2 = xy$ is then given by $x = u^2v$, $y = v$ and $z = uv$. Thus, as we already know, we resolve $z^2 = xy$ by M(2). Thus the weighted graph for $z^2 = xy$ is $\underset{-2}{\cdot}$ with the vertex the Riemann surface P^1.

$V = \{z^2 = x(x^2+y^2)\}$ has branch locus $B = \{x(x^2+y^2) = 0\}$. B has a singularity at the origin where the three planes $B_1 = \{x = 0\}$, $B_2 = \{x + iy = 0\}$, $B_3 = \{x - ix\}$ intersect.

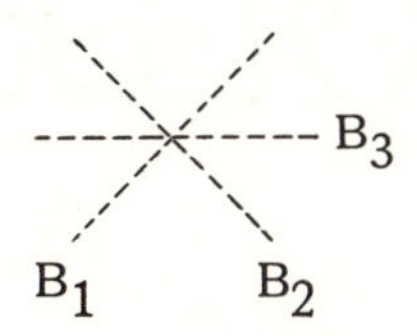

In order to put the branch locus into useful form, we must perform a quadratic transformation at $(x,y) = (0,0)$. Let $(x,y) = (\xi\zeta, \zeta) = (\zeta', \xi'\zeta')$ Then V' is given by $z^2 = \xi\zeta(\zeta^2 + \xi^2\zeta^2) = \xi\zeta^3(1+\xi^2)$ and $z^2 = \zeta'(\zeta'^2 + \xi'^2\zeta'^2) = \zeta'^3(1+\xi'^2)$. Let A_1 be the P^1 introduced by the quadratic transformation. $A_1 = \{\zeta = \zeta' = 0\}$. The new branch locus looks like

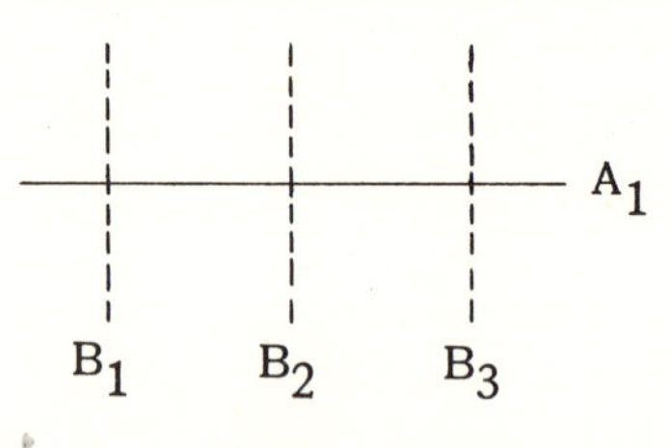

where we shall use dotted lines for branch curves which will not eventually appear in $\pi^{-1}(0)$. V' has singularities at $\xi = 0, \pm i$ and each singularity is locally the same as $z^2 = \xi\zeta^3$. The branching over the rest of the branch locus is just a connected branched two-fold cover.

We resolve $z^2 = \xi\zeta^3$ as before, first restricting to $|\xi| = |\zeta| = 1$. $(\xi,\zeta,z) = (e^{2\pi i\theta}, e^{2\pi i\phi}, e^{\pi i\theta + 3\pi i\phi})$ expresses $z^2 = \xi\zeta^3$ as the image of the universal covering space of $|\xi| = |\zeta| = 1$. This has the same lattice of $z^2 = xy$ so an equivalent covering is, as before, $(u,v) = e^{\pi i(\theta - \phi)}, e^{2\pi i\phi})$. The map from (u,v) to $z^2 = \xi\zeta^3$ is then $\zeta = v$, $\xi = u^2v$ and $z = uv^2$. The equations in the other coordinate system for $M(2)$ are $\zeta = u'^2v'$, $\xi = v'$ and $z = u'^3v'^2$. Let $A_2 = \{v = v' = 0\}$. The singularities at $\xi = \pm i$ yield A_3 and A_4. Thus, the weighted graph looks like

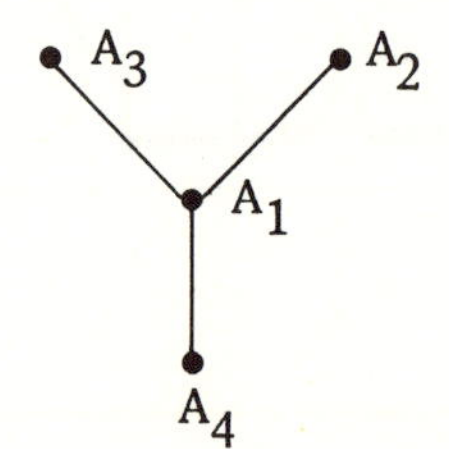

but we must still determine the weights. Actually we know that $A_2 \cdot A_2 = A_3 \cdot A_3 = A_4 \cdot A_4 = -2$, but in determining $A_1 \cdot A_1$ using Theorem 2.6 we will get all the weights.

Since z is an analytic function on V, it is certainly analytic on the resolution. Moreover its zero set, with order of vanishing, looks like:

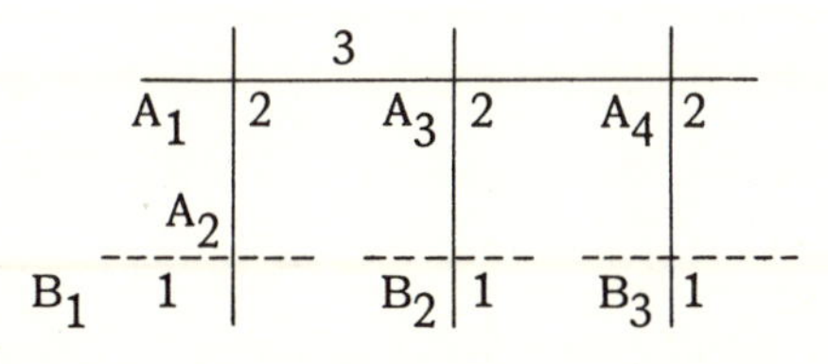

So $(z)\cdot A_1 = 0 = 3A_1\cdot A_1 -2-2-2$. Hence $A_1\cdot A_1 = -2$.
$(z)\cdot A_2 = 0 = 2A_2\cdot A_2 - 3-1$. So $A_2\cdot A_2 = -2$. Thus our weighted graph is

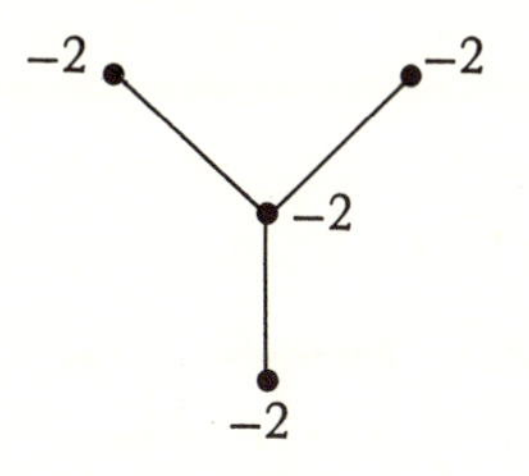

with each vertex a P^1.

$V = \{z^5 = x^2 + y^3\}$ is our next example. The branch locus $B = \{x^2 + y^3 = 0\}$ has a singularity at the origin.

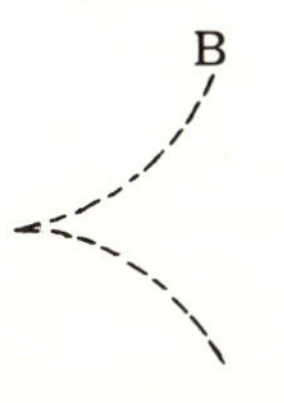

Let $(x,y) = (\zeta\xi, \zeta) = (\zeta', \xi'\zeta')$. V' is given by $z^5 = \zeta^2\xi^2 + \zeta^3 = \zeta^2(\xi^2 + \zeta)$ and $z^5 = \zeta'^2 + \xi'^3\zeta'^3 = \zeta'^2(1 + \xi'^3\zeta')$. Let $A_1 = \{\zeta = \zeta' = 0\}$. The two components of B', A_1 and $B = \{\xi^2 + \zeta = 0\}$ do not cross transversely at the origin so additional quadratic transformations are necessary.

Let $(\zeta,\xi) = (\sigma\tau, \tau) = (\tau', \sigma'\tau')$. $z^5 = \sigma^2\tau^2(\tau^2 + \sigma\tau) = \sigma^2\tau^3(\tau+\sigma)$ and $z^5 = \tau'^2(\sigma'^2\tau'^2 + \tau') = \tau'^3(\sigma'^2\tau' + 1)$. Let $A_2 = \{\tau = \tau' = 0\}$. A_1 lifts to $\sigma = 0$ and B lifts to $\tau + \sigma = 0$. Hence the new branch locus looks like

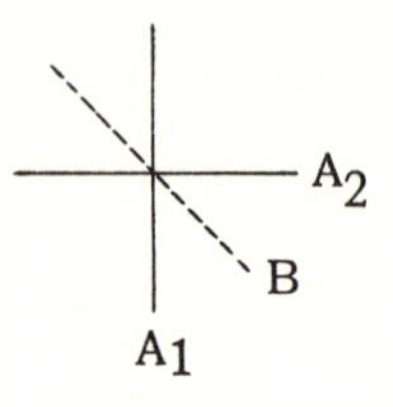

and one more quadratic transformation is needed. Let $(\sigma,\tau) = (st, t) = (t', s't')$. $z^5 = s^2t^2t^3(st+t) = s^2t^6(s+1)$ and $z^5 = t'^2s'^3t'^3(t'+s't') = s'^3t'^6(s'+1)$. Let $A_3 = \{t = t' = 0\}$. The new branch locus looks like:

$A_1 = \{s = 0\}$ $A_2 = \{s' = 0\}$

$A_3 = \{t = t' = 0\}$

$B = \{s = -1\}$

Where the branch locus is non-singular we have a 5-fold connected analytic cover. At $s = 0$, the singularity is equivalent to $z^5 = s^2t^6$. Restricting as usual to $|s| = |t| = 1$, $(s,t,z) = (e^{2\pi i\phi}, e^{\frac{4\pi i\theta + 12\pi i\phi}{5}})$ expresses $z^5 = s^2t^6$ as the image of its universal covering space. The associated lattice is

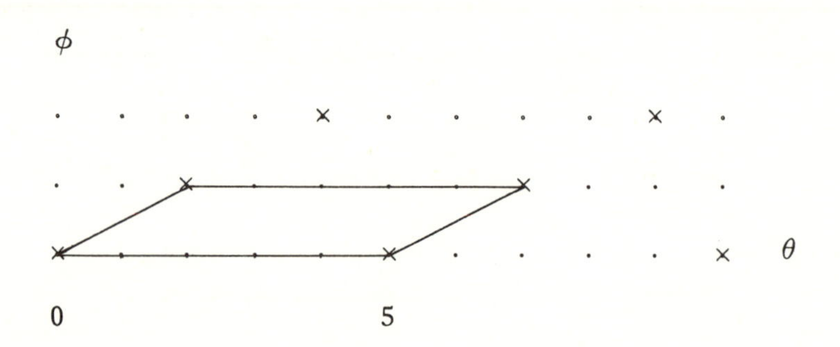

A fundamental parallelogram may be chosen with vertices (0,0), (5,0), (2,1) and (7,1). An equivalent covering may is thus given by $(u,v) = (e^{\frac{2\pi i(\theta-2\phi)}{5}}, e^{2\pi i\phi})$. Then $(s,t,z) = (u^5v^2, v, u^2v^2)$. To find $M(k_1, \ldots, k_s)$, express $\frac{5}{2}$ as a continued fraction $5/2 = 3 - \frac{1}{2}$. Thus $k_1 = 3$ and $k_2 = 2$ and the resolution in the other coordinate systems is $(s,t,z) = (u'v'^2, u'^3v', u'^4v'^2) = (u'', u''^3v''^5, u''^4v''^6)$. A_1 lifts to $u = 0$ and A_3 lifts to $v'' = 0$.

At $s' = 0$, the singularity is equivalent to $z^5 = s'^3t'^6$. $(s',t',z') = (e^{2\pi i\theta}, e^{\pi i\phi}, e^{\frac{6\pi i\theta + 12\pi i\phi}{5}})$. The lattice is

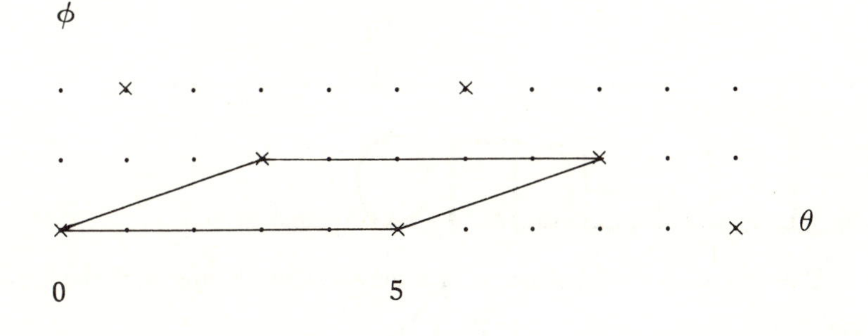

$(u,v) = (e^{\frac{2\pi i(\theta-3\phi)}{5}}, e^{2\pi i\phi})$. $(s',t',z) = (u^5v^3, v, u^3v^3)$. $5/3 = 2 - \frac{1}{3}$ so we use $M(2,3)$. In its other coordinate patches, $(s',t',z) = (u'v'^3, u'^3v', u'^3v'^3) = (u'', u''^2v''^5, u''^3v''^6)$. A_2 lifts to $u = 0$ and A_3 lifts to $v'' = 0$.

Finally at $s = -1$, the singularity is $z^5 = (s+1)t^6$. $(s+1,t,z) = (e^{2\pi i\theta}, e^{2\pi i\phi}, e^{\frac{2\pi i\theta + 12\pi i\phi}{5}})$. The lattice is

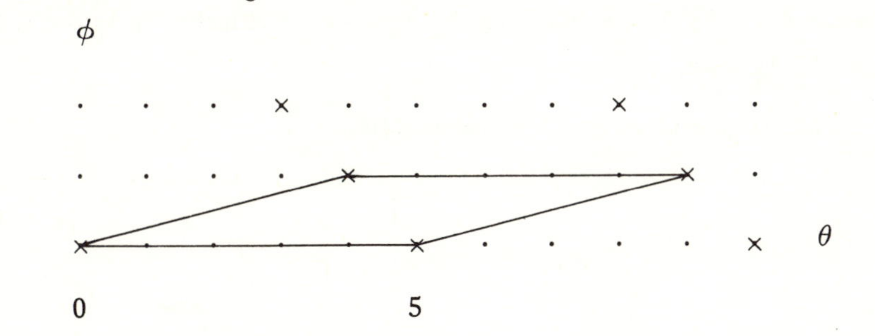

$(u,v) = (e^{\frac{2\pi i(\theta-4\phi)}{5}}, e^{2\pi i\phi})$. $(s+1,t,z) = (u^5v^4, v, uv^2)$.

$$5/4 = 2 - 3/4 = 2 - \frac{1}{4/3} = 2 - \frac{1}{2 - 2/3} = 2 - \frac{1}{2} - \frac{1}{3/2} = 2 - \cfrac{1}{2 - \cfrac{1}{2 - \cfrac{1}{2}}}.$$

Hence we introduce $M(2,2,2,2)$. A_3 lifts to $v^{(4)} = 0$ and B lifts to $u = 0$. $z = uv^2 = u'^3v'^2 = u''^3v''^4 = u'''^5v'''^4 = (u^{(4)})^5(v^{(4)})^6$.

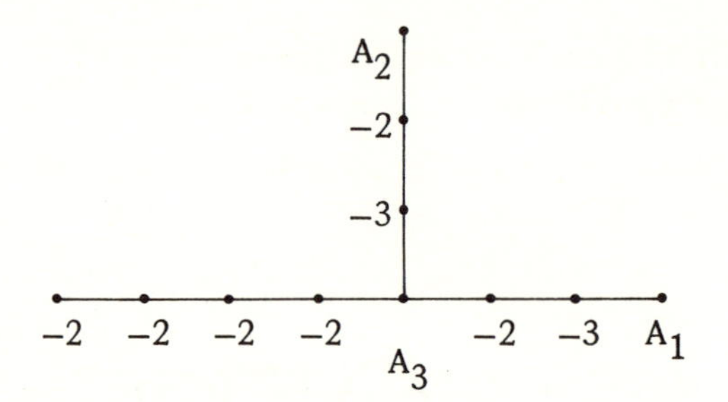

and we must still determine $A_3 \cdot A_3$, $A_2 \cdot A_2$ and $A_1 \cdot A_1$.

Use the analytic function z. Its zero set, with orders of the zeros looks like

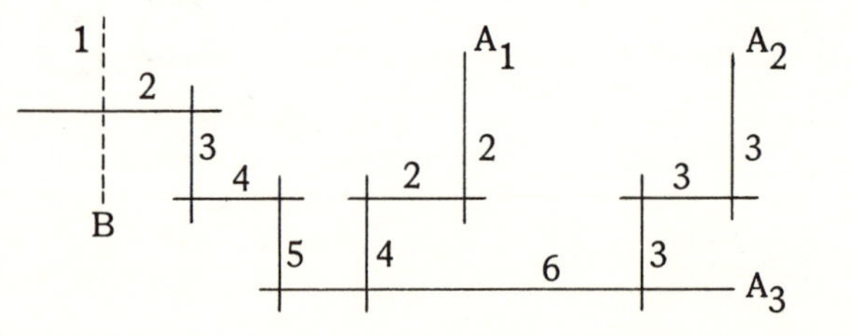

Hence $0 = 6\,A_3 \cdot A_3 + 5 + 4 + 3$ so $A_3 \cdot A_3 = -2$. Similarly $A_2 \cdot A_2 = -1$, and $A_1 \cdot A_1 = -1$.

Thus the weighted graph of our resolution is

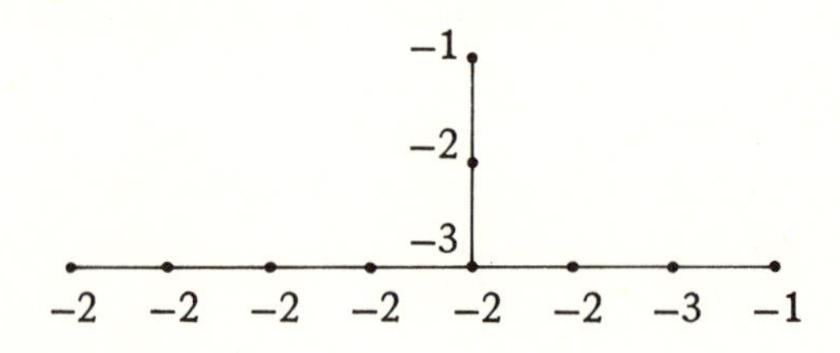

where each vertex is a P^1.

This is not the usual weighted graph for $z^5 = x^2 + y^3$ so a few words are in order. As we shall see later, Theorem 5.5, any P^1 with self-intersection number -1 is the result of a quadratic transformation and thus may be collapsed to a point without creating any singularity. This collapsing increases by $+1$ the self-intersection number of any submanifold originally meeting the P^1 transversely, Lemma 4.3. Thus, collapsing A_1 and A_2, we get a weighted graph

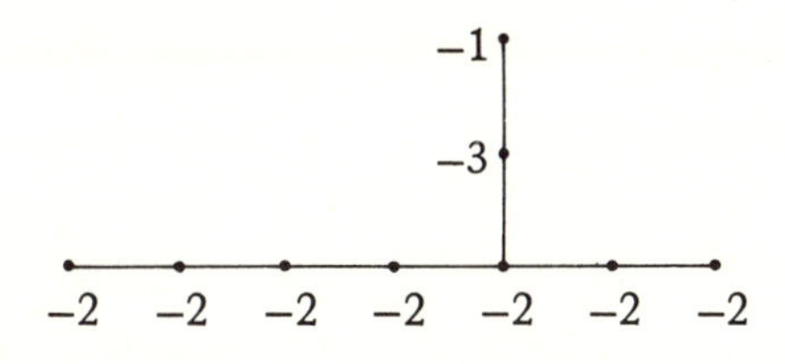

This in turn yields

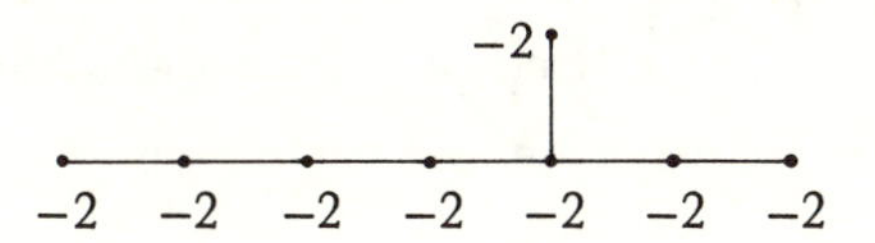

which may be recognized as the Dynkin diagram for the Lie group E_8.

Our next example is $V = \{z^6 = x^2 + y^3\}$. This has the same branch locus as our previous example, so the first calculations are exactly the same as before. The final equations are $V' = \{z^6 = s^2t^6(s+1)\}$ and $V' = \{z^6 = s'^3t'^6(s'+1)\}$. However V' is not locally irreducible as were all of our previous examples. Above A_1, for $s = 0$, $t \neq 0$, the singularity of V' is equivalent to $z^6 = s^2$, or $(z^3-s)(z^3+s) = 0$. So the analytic cover consists of two 3-fold branched covers. Above A_2, for $s' = 0$, $t' \neq 0$, we have three irreducible components, each a two-fold cover. Above A_3 for $t = 0$, $s \neq 0, 1$ we have six non-singular irreducible components.

At $s = t = 0$, we must resolve a singularity equivalent to $z^6 = s^2t^6$ or $(z^3-st^3)(z^3+st^3) = 0$. Thus we have two irreducible components, each with the same singularity. The universal covering map for $z^3 = st^3$ for $|s| = |t| = |z| = 1$ is $(s,t,z) = (e^{2\pi i\theta}, e^{2\pi i\phi}, e^{\frac{2\pi i\theta + 6\pi i\phi}{3}})$. The lattice is

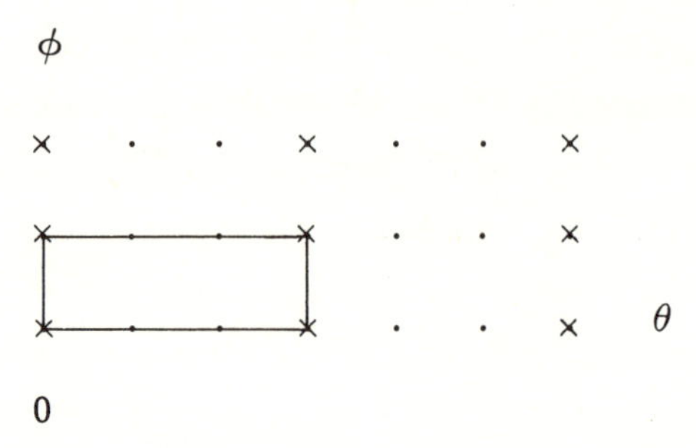

$(u,v) = (e^{\frac{2\pi i\theta}{3}}, e^{2\pi i\theta})$. $(s,t,z) = (u^3, v, uv)$ which gives $M(\emptyset)$ as the resolving manifold, i.e. we have introduced no new P^1's.

At $s' = 0$, the singularity is $z^6 = s'^3t'^6$ which has three irreducible components, each equivalent to $z^2 = s't'^2$. This is resolved by $M(\emptyset)$, $(s',t',z) = (u^2, v, uv)$. Finally at $s = -1$, the singularity is $z^6 = (s+1)t^6$. This is irreducible and resolved by $M(\emptyset)$, $(s+1,t,z) = (u^6, v, uv)$.

We must still determine the weighted graph. We have the projection map $\pi: M \to V'$ with $\pi^{-1}(A_1 \cup A_2 \cup A_3)$ being the inverse image of the origin in $z^6 = x^2 + y^3$. Over A_3, we have a 6-fold cover with branch points at $s = 0$, $s = -1$ and $s' = 0$. At $s = 0$, we have two 3-fold branch points. At $s' = 0$, we have two 3-fold branch points. At $s = -1$, we have one 6-fold branch point, which shows that $\pi^{-1}(A_3) = \tilde{A}_3$ is connected. We can determine the genus of $\tilde{A}_3$ using χ, the Euler characteristic, $\chi = 2-2g$. If α, β and γ are respectively the number of 0,1 and 2-simplices in a triangulation of A_3, then $\alpha - \beta + \gamma = 2$. We may choose a triangulation such that each branch point is a 0-simplex. On $\tilde{A}_3$, we have 6γ 2-simplices, 6β 1-simplices and $6\alpha - 4 - 3 - 5$ 0-simplices. Hence $\chi(\tilde{A}_3) = 6\alpha - 12 - 6\beta + 6\gamma = 6(\alpha - \beta + \gamma) - 12 = 12-12 = 0$. Hence $\tilde{A}_3$ is a torus.

Over A_1 we have a 2-fold cover, with $t' = 0$ the only possible branch point. $t' = 0$ is in fact not a branch point, so $\pi^{-1}(A_2) = A_{11} \cup A_{12}$ where A_{11} and A_{12} are projective lines which do not intersect. Similarly, $\pi^{-1}(A_2) = A_{21} \cup A_{22} \cup A_{23}$ where the A_{21} are non-intersecting projective lines. Thus our graph looks like:

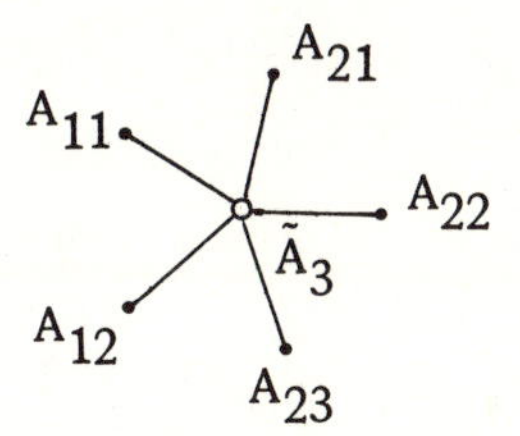

and we must determine the weights.

Use the analytic function Z. Its zero set, with order of the zeros, looks like

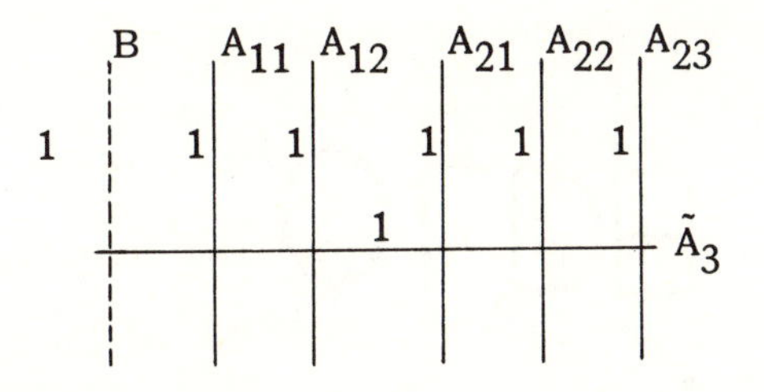

Thus the weighted graph is

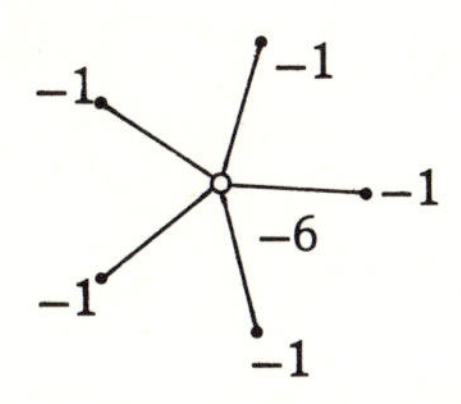

with the center vertex a torus and the other vertices P^1's. To get the usual resolution of $z^6 = x^2 + y^3$, we collapse, as mentioned before, the five projective lines with self-intersection -1. This gives a new resolution with weighted graph

$$\circ \; -1$$

and the vertex is a torus.

Thus it is possible for $\pi^{-1}(p)$, where p is a singular point to have irreducible components which are not projective lines. Our next and last example shows that it is possible for the graph to contain cycles. We will eventually obtain a complete characterization of all possible weighted graphs.

Consider finally $V = \{z^2 = (x+y^2)(x^2+y^7)\}$ which has an isolated singularity at the origin. $B_1 = \{x+y^2 = 0\}$ and $B^2 = \{x^2+y^7 = 0\}$ are the two irreducible components of the branch locus. Performing successive quadratic transformations, $(x,y) = (\xi\zeta,\zeta) = (\zeta',\xi'\zeta')$.
$z^2 = (\xi\zeta+\zeta^2)(\xi^2\zeta^2+\zeta^7) = \zeta^3(\xi+\zeta)(\xi^2+\zeta^5)$ and
$z^2 = (\zeta'+\xi'^2\zeta'^2)(\zeta'^2+\zeta'^7\zeta'^7) = \zeta'^3(1+\xi'^2\zeta)(1+\xi'^7\zeta'^5)$.
$A_1 = \{\zeta = \zeta' = 0\}$. The branch locus is

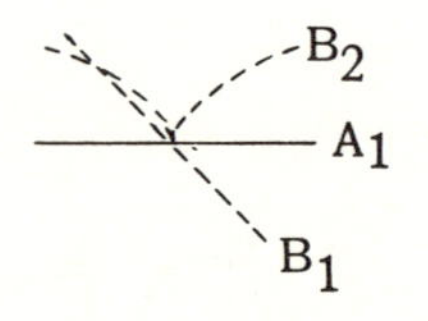

$(\xi,\zeta) = (\sigma\tau,\tau) = (\tau',\sigma'\tau')$. $z^2 = \tau^3(\sigma\tau+\tau)(\sigma^2\tau^2+\tau^5) = \tau^6(\sigma+1)(\sigma^2+\tau^3)$ and
$z^2 = \sigma'^3\tau'^3(\tau'+\sigma'\tau')(\tau'^2+\sigma'^5\tau'^5) = \sigma'^3\tau'^6(1+\sigma')(1+\sigma'^5\tau'^3)$.
$A_2 = \{\tau = \tau' = 0\}$.

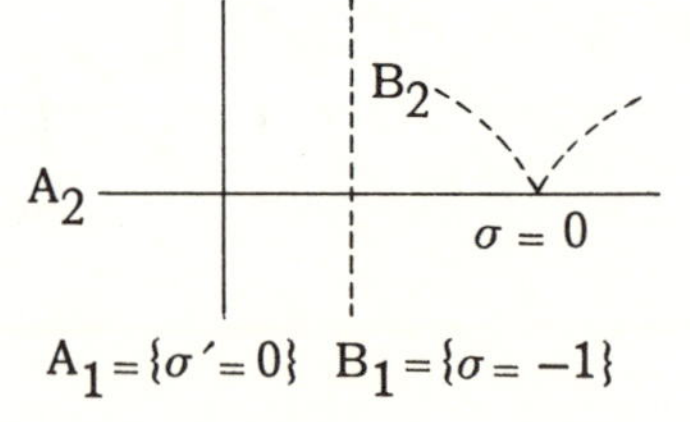

The singularity at $\sigma = 0$ is equivalent to $z^2 = \tau^6(\sigma^2+\tau^3)$.
$(\sigma,\tau) = (st,t) = (t',s't')$. $z^2 = t^6(s^2t^2+t^3) = t^8(s^2+t)$.
$z^2 = s'^6t'^6(t'^2+s'^3t'^3) = s'^6t'^8(1+s'^3t')$. $A_3 = \{t = t' = 0\}$.

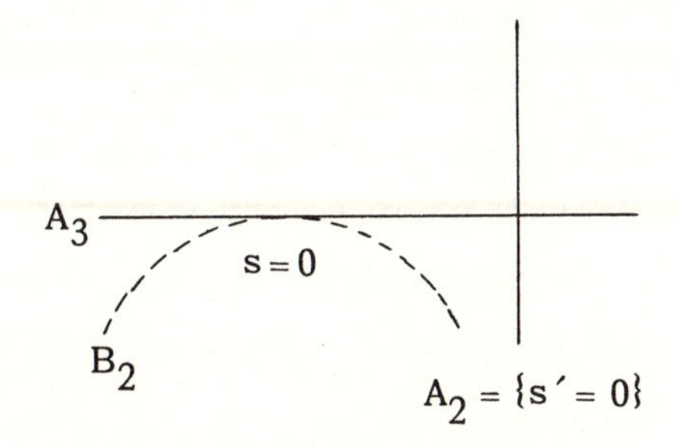

$(s,t) = (\alpha\beta, \beta) = (\beta', \alpha'\beta')$. $z^2 = \beta^8(\alpha^2\beta^2 + \beta) = \beta^9(\alpha^2\beta + 1)$.
$z^2 = \alpha'^8\beta'^8(\beta'^2 + \alpha'\beta') = \alpha'^8\beta'^9(\beta' + \alpha')$. $A_4 = \{\beta = \beta' = 0\}$.

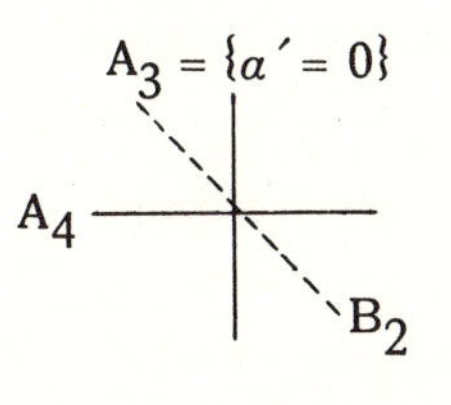

One last transformation! $(\alpha', \beta') = (qr, r) = (r', q'r')$.
$z^2 = q^8r^8r^9(qr + r) = q^8r^{18}(q+1)$ and $z^2 = r'^8q'^9r'^9(r' + q'r') = q'^9r'^{18}(1 + q')$.
$A_5 = \{r = r' = 0\}$.

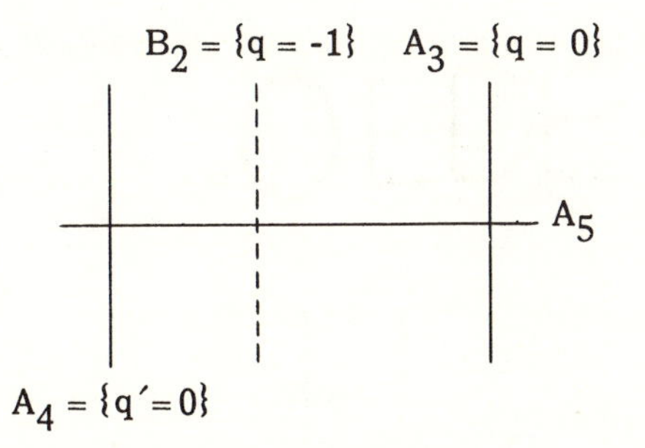

The singularity at $A_2 \cap B_1$ is equivalent to $z^2 = \tau^6(\sigma + 1)$ which is resolved by $M(\emptyset)$; $(\sigma + 1) = u^2$, $\tau = v$, $z = uv^3$. The singularity at $A_2 \cap A_1$ is equivalent to $z^2 = \sigma'^3\tau'^6$ which is resolved by $M(\emptyset)$; $\sigma' = u^2$, $\tau' = v$, $z = u^3v^3$. The singularity at $A_3 \cap A_2$ is equivalent to $z^2 = s'^6t'^8$

which has two irreducible components, each of which is non-singular. The singularity at $A_5 \cap A_4$ is equivalent to $z^2 = q'^9 r'^{18}$ which is resolved by $M(\emptyset)$; $q' = u^2$, $r' = v$, $z = u^9 v^9$. The singularity at $A_5 \cap B_2$ is equivalent to $z^2 = r^{18}(q+1)$ which is resolved by $M(\emptyset)$; $(q+1) = u^2$, $r = v$, $z = uv^9$. Finally, the singularity at $A_5 \cap A_3$ is equivalent to $z^2 = q^8 r^{18}$ which has two non-singular irreducible components.

If, as in the previous example, we let $\pi : M \to V'$ be the projection map, π establishes a 1-fold cover over A_4 and over A_1. There is a 2-fold cover over A_2 with $\sigma = 0$ and $\sigma = -1$ branch points. Thus $\tilde{A}_2 = \pi^{-1}(A_2)$ is just a projective line. Hence $\pi^{-1}(A_3) = A_{31} \cup A_{32}$ with A_{31} and A_{32} projective lines which do not intersect. Finally, there is a 2-fold cover of A_5, branched over $q = 0$ and $q = 1$, so $\tilde{A}_5 = \pi^{-1}(A_5)$ is a $\mathbf{P}^1$. This gives a graph

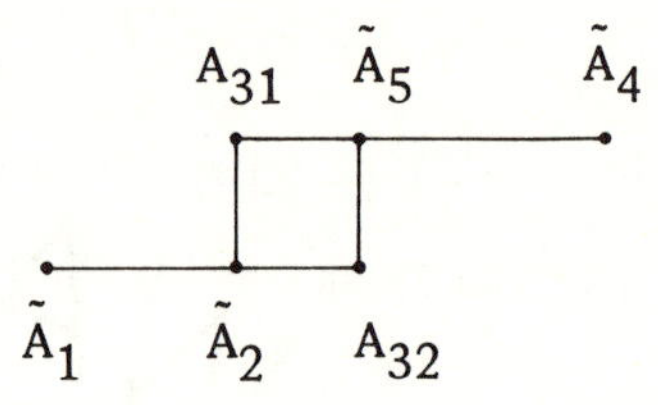

We determine the weights by looking, as usual, at the zeros with multiplicities of z.

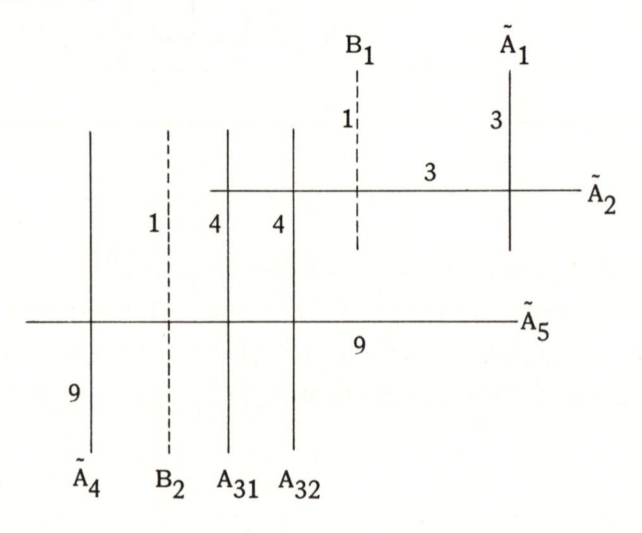

Thus the weighted graph is

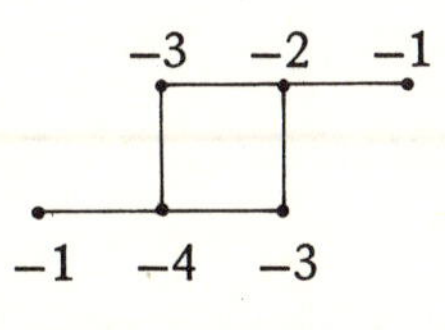

Collapsing $\tilde{A}_1$ and $\tilde{A}_4$ gives

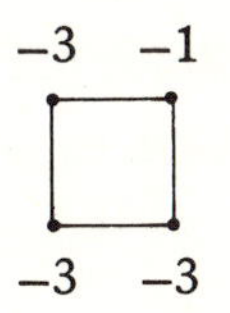

Finally we may collapse the vertex with weight -1. The two curves originally meeting this vertex now meet each other. As usual, their weights are increased by 1. Thus there is a resolution of V with weighted graph

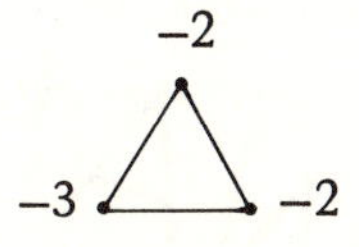

and each vertex is a projective line.

CHAPTER III

NORMALIZATION OF TWO-DIMENSIONAL ANALYTIC SPACES

As observed in the case of plane curves and at all but a discrete set of points in the 2-dimensional case, the local resolution of a singularity consisted of taking irreducible components of the variety and then, without changing into the underlying topological space, making the irreducible components into complex manifolds by adding more holomorphic functions. This process of, roughly, allowing the analytic space to have as many holomorphic functions as possible is called normalization. Formally,

DEFINITION 3.1. *Let* V *be an analytic space,* $a \in V$. *A germ* h *of a function defined on the regular points of* V *near* a *is said to be weakly holomorphic at* a *if* h *is holomorphic on the regular points near* a *and locally bounded near* a. *Let* $\tilde{\mathcal{O}}$ *be the sheaf of germs of weakly holomorphic functions. There is a natural inclusion* $\mathcal{O} \subset \tilde{\mathcal{O}}$. V *is normal at* a *if* $\mathcal{O}_a \subset \tilde{\mathcal{O}}_a$ *is an isomorphism.* V *is normal if* $\mathcal{O} \approx \tilde{\mathcal{O}}$, *i.e if* V *is normal at each of its points.*

Certainly every manifold point is normal. If V_a is reducible, then V cannot be normal at a, for let $h \equiv 1$ on one component of the regular set and let $h \equiv 0$ on the other components. Theorem III. C. 19 of G & R implies that the number of components of the regular set near a equals the number of irreducible components of V_a. h is weakly holomorphic but cannot be holomorphic since it cannot be continuous at a.

The following theorem is useful in verifying that certain singular points are normal. In particular, all of the singularities resolved in section II were normal.

THEOREM 3.1 *Let* V *be a purely 2-dimensional subvariety in the 3-dimensional manifold* M. *If* p *is an isolated singularity of* V, *then* V *is normal at* p.

Proof: We shall in fact prove that any function h holomorphic in a punctured neighborhood of p has a holomorphic extension to p. Choose Δ, a polydisc neighborhood of p in some coordinate system, with $p = (0,0,0)$ the only singularity of V in Δ. Let $\mathcal{J}$ be the ideal sheaf of V. We have the exact sheaf sequence

$$0 \to \mathcal{J} \to \mathcal{O} \to {}_V\mathcal{O} \to 0 \tag{3.1}$$

where ${}_V\mathcal{O}$ is the sheaf of germs of holomorphic functions on V. (3.1) yields the usual exact long cohomology sequence. It thus suffices to show that $H^1(\Delta\text{-p}, \mathcal{J}) = 0$, for then every holomorphic function $h \in \Gamma((\Delta \cap V) - p, {}_V\mathcal{O}) = \Gamma(\Delta\text{-p}, {}_V\mathcal{O})$ will be the restriction of a holomorphic function $\tilde{h} \in \Gamma(\Delta\text{-p}, \mathcal{O})$. $\tilde{h}$ extends to be holomorphic in Δ by Hartogs' theorem and then restricts back to V to give a holomorphic extension of h.

By Lemma VIII. B.12 of G & R, there is a holomorphic function f in Δ such that f_x generates $\mathcal{J}_x$ at each point $x \in \Delta$. Thus the map $f\colon \mathcal{O} \to \mathcal{J}$ given by $f(g) = f \cdot g$ establishes a sheaf isomorphism $\mathcal{O} \approx \mathcal{J}$. Hence $H^1(\Delta\text{-p}, \mathcal{J}) \approx H^1(\Delta\text{-p}, \mathcal{O})$. We compute $H^1(\Delta\text{-p}, \mathcal{O})$ by a Leray covering.

$$\Delta\text{-p} = \{(x,y,z) \mid (x,y,z) \neq (0,0,0), |x| < 1, |y| < 1, |z| < 1\}.$$

Our Leray covering will have three open sets, $U_1 = \{x \neq 0\}$, $U_2 = \{y \neq 0\}$, and $U_3 = \{z \neq 0\}$. There are three 1-simplices, $U_1 \cap U_2 = \{x \neq 0, y \neq 0\}$, $U_2 \cap U_3 = \{y \neq 0, z \neq 0\}$ and $U_3 = \{z \neq 0, x \neq 0\}$. There is one 3-simplex, $U_1 \cap U_2 \cap U_3 = \{x \neq 0, y \neq 0, z \neq 0\}$. Suppose

$$g = (g_1 \in \Gamma(U_1 \cap U_2, \mathcal{O}), g_2 \in \Gamma(U_2 \cap U_3, \mathcal{O}), g_3 \in \Gamma(U_3 \cap U_1, \mathcal{O}))$$

is a 1-cochain. We may expand the g_i in unique Laurent series:

$$g_1 = \sum_{\substack{-\infty < \nu_1 < \infty \\ -\infty < \nu_2 < \infty \\ 0 \leq \nu_3 < \infty}} a_{\nu_1\nu_2\nu_3} x^{\nu_1} y^{\nu_2} z^{\nu_3}$$

$$g_2 = \sum_{\substack{0 \leq \nu_1 < \infty \\ -\infty < \nu_2 < -\infty \\ -\infty < \nu_3 < -\infty}} b_{\nu_1\nu_2\nu_3} x^{\nu_1} y^{\nu_2} z^{\nu_3}$$

$$g_3 = \sum_{\substack{-\infty < \nu_1 < \infty \\ 0 \leq \nu_2 < \infty \\ -\infty < \nu_1 < -\infty}} c_{\nu_1\nu_2\nu_3} x^{\nu_1} y^{\nu_2} x^{\nu_3}$$

If g is a cocycle, i.e. $\delta g = 0$, then $a_{\nu_1\nu_2\nu_3} = 0$ if $\nu_1 < 0$ and $\nu_2 < 0$. Hence $g_1 = f_1 - f_2$ with

$$f_1 = \sum_{\substack{-\infty < \nu_1 < \infty \\ 0 \leq \nu_2 < \infty \\ 0 \leq \nu_3 < \infty}} a_{\nu_1\nu_2\nu_3} x^{\nu_1} y^{\nu_2} z^{\nu_3} \in \Gamma(U_1, \mathcal{O})$$

and

$$f_2 = \sum_{\substack{0 \le \nu_1 < \infty \\ -\infty < \nu_2 < 0 \\ 0 \le \nu_3 < \infty}} -a_{\nu_1\nu_2\nu_3} x^{\nu_1} y^{\nu_2} z^{\nu_3} \in \Gamma(U_2, \mathcal{O})$$

It is now easy to use the rest of the cocycle condition to compute f_3 and show that g is a coboundary.▮

The next very general lemma will prove useful.

LEMMA 3.2. *Let* $\pi: Y \to V$ *be a proper surjective holomorphic map between the analytic spaces* Y *and* V. *Let* $U_1, \ldots, U_k$ *be neighborhoods of the connected components of* $\pi^{-1}(p)$, $p \in V$. *Then for all sufficiently small neighborhoods* N *of* p, $\pi^{-1}(N) \subset \cup U_1$.

Proof: Suppose not, then there would exist a sequence $\{p_j\} \to p$ with $\pi^{-1}(p_j) \not\subset \cup U_i$. $\pi^{-1}(p,\{p_j\})$ is compact, so there would exist a subsequence $\{p'_{j'}\} \in Y$ with $p'_{j'} \notin \cup U_i$ and $\pi(p'_{j'}) = p_{j'}$ with $p'_{j'} \to p'$, $\pi(p') = p$. But $p' \in \cup U_i$, a contradiction.▮

DEFINITION 3.2. *Let* V *be an analytic space. A normalization* (Y,π) *of* V *is a normal analytic space* Y *and a holomorphic map* $\pi: Y \to V$ *such that*

(i) $\pi: Y \to V$ *is proper and has finite fibres.*

(ii) *If* S *is the singular set of* V *and* $A = \pi^{-1}(S)$, *then* $Y-A$ *is dense in* Y *and* $\pi|Y-A$ *is biholomorphic.*

PROPOSITION 3.3. *If* $p \in V$ *and* (Y,π) *is a normalization of* V, *then the number of points in* $\pi^{-1}(p)$ *equals the number of irreducible components of* V_p.

Proof: Let $\{q_1, \ldots, q_r\} = \pi^{-1}(p)$. Let $U_1, \ldots, U_r$ be disjoint neighborhoods of $q_1, \ldots, q_r$ respectively. Let N be a neighborhood of p with $\pi^{-1}(N) \subset \cup U_i$. Theorem II. C. 19 of G & R implies that the number of components of V–S near a equals the number of irreducible components of V_a. Then V_a has at least r components. On the other hand, Y is irreducible so the regular points R_i in U_i are connected near q_i. A is a nowhere dense subvariety of Y. Hence by Corollary I. C. 4 of G&R, $R_i - A = U_i \cap (Y-A) = U_i \cap \pi^{-1}(V-S)$ is connected near a. π is biholomorphic on Y–A so V_a has exactly r components.∎

THEOREM 3.4. *Let* (Y,π) *and* (Y',π') *be normalizations of* V. *Then there is a unique biholomorphic map* $\rho\colon Y \to Y'$ *such that* $\pi' \circ \rho = \pi$.

Proof: On Y–A, $\rho = (\pi')^{-1} \circ \pi$. Hence ρ is determined on a dense set and must necessarily be unique. We must show that ρ extends to a holomorphic map. Let $a \in A$. Let $U_1, \ldots, U_k$ be neighborhoods of the points of $(\pi')^{-1}\pi(a)$ such that the disjoint union of the U_i may be embedded in a bounded (disconnected) region in $\mathbb{C}^n$, some n. Let N be a neighborhood of $\pi(a)$ as in Lemma 3.2. In $\pi^{-1}(N)-A$, and in particular near a, the mapping ρ is given by bounded holomorphic functions. Since A is a proper subvariety of Y, ρ extends through any regular points in A. Since Y is normal, ρ extends to a holomorphic map.∎

In view of Theorem 3.4, we may speak of *the* normalization of V.

THEOREM 3.5. *Any purely 2-dimensional analytic space* V *has a normalization.*

The proof of Theorem 3.5 will require a fair amount of preliminary work and is postponed until after Lemma 3.11. In fact any analytic space has a normalization.

DEFINITION 3.3. *Let* V_a *be a variety. A universal denominator* u *for* V_a *is a germ of an analytic function at* a *such that* u *is not a zero divisor, and such that given any germ* h *of a weakly holomorphic function at* a, u h *may be extended to the germ of a holomorphic function at* a.

Theorems III. B. 21 and III. B. 23 of G & R insure the existence of a universal denominator u if V_a is of pure dimension. u is in fact a universal denominator in some neighborhood of a.

As before, let $\tilde{\mathcal{O}}$ be the sheaf of germs of weakly holomorphic functions.

THEOREM 3.6. *If* v *is pure dimensional,* $\tilde{\mathcal{O}}_a$ *is the integral closure of* $\mathcal{O}_a$ *in the total quotient ring.*

Proof: If h is weakly holomorphic at a, then $h = \frac{g}{u}$ where g is holomorphic and u is a universal denominator. Hence h is in the total quotient ring. h is integral over $\mathcal{O}_a$ means that h satisfies a monic polynomial with integral coefficients. This is just Corollary III. B. 14 of G & R.

Conversely if $\frac{f}{g}$, g not a zero divisor, is integral, i.e.

$$(\frac{f}{g})^{\lambda} + a_1(\frac{f}{g})^{\lambda-1} + \ldots + a_{\lambda} \equiv 0 \tag{3.2}$$

where the a_i are holomorphic functions, we must show that $\frac{f}{g}$ is weakly holomorphic. g cannot vanish identically on any connected component of the regular set, for suppose that g did. Let h be identically 1 on this component and otherwise 0. u h is holomorphic and not identically 0. $(uh)g \equiv 0$ implies that g is a zero divisor, a contradiction. It thus suffices to show that $\frac{f}{g}$ is locally bounded, for the Riemann removable singularity theorem then guarantees that $\frac{f}{g}$ may be extended holomorphically through $g = 0$ to all the regular points of V.

Pick a neighborhood of a where the a_i in (3.2) have a common bound. The roots of a polynomial with leading coefficient 1 are bounded in terms of the degree and the other coefficients. Hence $\frac{f}{g}$ is bounded. ▮

THEOREM 3.7. *Let* V *be a pure dimensional analytic space. Then the set* N *of* $p \in V$ *such that* V *is not normal at* p *is an analytic subvariety of* V.

Proof: [Grauert-Remmert] If $p \in V$, it suffices to prove that N is a subvariety near p. Let U be a neighborhood of p where we have a universal denominator u. Let $S = \{a \in U | u(a) = 0\}$. Let $\mathcal{I}$ be the sheaf of germs of holomorphic functions on V which vanish on S. Let us verify that $\mathcal{I}$ is a coherent sheaf of ${}_V\mathcal{O}$-modules. Locally $S \subset V \subset \Delta \subset \mathbf{C}^n$, some n. Let ${}_S\mathcal{I}$ and ${}_V\mathcal{I}$ be the ideal sheaves for S and V as ambient $\mathcal{O}$-modules. The canonical map $\mathcal{O} \to {}_V\mathcal{O} = \mathcal{O}/{}_V\mathcal{I}$ gives any ${}_V\mathcal{O}$-module an $\mathcal{O}$-module structure. ${}_S\mathcal{I}$ and ${}_V\mathcal{I}$ are coherent sheaves of $\mathcal{O}$-modules. Hence ${}_S\mathcal{I}/{}_V\mathcal{I}$ is a coherent sheaf of $\mathcal{O}$-modules. It is easy to verify that the $\mathcal{O}$-module structure on ${}_S\mathcal{I}/{}_V\mathcal{I}$ is isomorphic to the $\mathcal{O}$-module structure on $\mathcal{I}$. Resolve ${}_S\mathcal{I}/{}_V\mathcal{I}$ by free $\mathcal{O}$-modules.

$$\mathcal{O}^m \overset{\alpha}{\to} \mathcal{O}^n \overset{\beta}{\to} {}_S\mathcal{I}/{}_V\mathcal{I} \to 0 \tag{3.3}$$

In (3.3) restrict the functions to V, i.e. take quotients with ${}_V\mathcal{I}$. This yields a sequence of ${}_V\mathcal{O}$-module homomorphisms

$${}_V\mathcal{O}^m \xrightarrow{\alpha_*} {}_V\mathcal{O}^n \xrightarrow{\beta_*} \mathcal{I} \to 0. \tag{3.4}$$

We must verify that (3.4) is exact. $\alpha_*\beta_* = 0$. To see that β_* is onto, suppose $g \in \mathcal{I}_x$. $g = \beta(f)$ for some $f \in \mathcal{O}^n_x$. Hence $g = \beta_*(\tilde{f})$ where $\tilde{f}$ is the restriction of f to V. If $\beta_*(\tilde{f}) = 0$, extend $\tilde{f}$ to $f \in \mathcal{O}^n_x$. $\beta(f)$ vanishes on V, and hence is trival in ${}_S\mathcal{I}/{}_V\mathcal{I}$. Hence $f = \alpha(h)$, or $\tilde{f} = \alpha_*(\tilde{h})$.

$\mathcal{H}om(\mathcal{I}, \mathcal{I})$ is coherent since, as we shall verify below, in general $\mathcal{H}om(\mathcal{R}, \mathcal{S})$ is coherent if $\mathcal{R}$ and $\mathcal{S}$ are coherent sheaves.

Let us recall that for arbitrary sheaves $\mathcal{R}$ and $\mathcal{S}$, $\mathcal{H}om(\mathcal{R}, \mathcal{S})$ is not defined stalkwise but is the sheaf of sheaf homomorphisms, i.e. $\Gamma(N, \mathcal{H}om(\mathcal{R}, \mathcal{S})) = \{\phi : \mathcal{R}|_N \to \mathcal{S}|_N$ such that ϕ is a sheaf homomorphism$\}$. However if $\mathcal{R}$ is coherent, everything may be done stalkwise, i.e.

PROPOSITION 3.8. *If $\mathcal{R}$ and $\mathcal{S}$ are sheaves of ${}_V\mathcal{O}$-modules and $\mathcal{R}$ is coherent, then $\mathcal{H}om_{{}_V\mathcal{O}}(\mathcal{R}, \mathcal{S})_x \approx \mathcal{H}om_{{}_V\mathcal{O}_x}(\mathcal{R}_x, \mathcal{S}_x)$ for all $x \in V$.*

Proof: There is a natural map ρ: $\mathcal{H}om_{_V\mathcal{O}}(\mathcal{R}, \mathcal{S})_x \to \mathrm{Hom}_{_V\mathcal{O}_x}(\mathcal{R}_x, \mathcal{S}_x)$ given by $(\rho(\phi))(r) = \phi_x(r)$. We must show that ρ is bijective. Let

$$\mathcal{O}_V^m \overset{\alpha}{\to} \mathcal{O}_V^n \overset{\beta}{\to} \mathcal{R} \to 0 \tag{3.5}$$

be a local resolution of $\mathcal{R}$.

Suppose $\rho\phi = 0$, i.e. $\phi\colon \mathcal{R}|_N \to \mathcal{S}|_N$ has $\rho_x\colon \mathcal{R}_x \to \mathcal{S}_x$ the trivial map. The coherence of $\mathcal{R}$ insures that the $r_1 = \beta(1,0,\ldots,0)$, $r_2 = \beta(0,1,\ldots,0),\ldots$ which generate $\mathcal{R}_x$ also generate $\mathcal{R}_y$ in some neighborhood of x. Hence upon restricting ϕ to such nearby y, ϕ becomes the trivial map. Hence ρ is injective.

Given $\psi\colon \mathcal{R}_x \to \mathcal{S}_x$, we must extend ψ to a sheaf map in some neighborhood of x. $\psi(r_1),\ldots,\psi(r_n)$ can all be represented by sections $s_1,\ldots,s_n$ of $\mathcal{S}$ in some common neighborhood of x. Given any $r \in \mathcal{R}_y$, $r = \Sigma\, d_i r_i$, $d_i \in {_V\mathcal{O}_y}$ and we let $\psi(r) = \Sigma\, d_i s_i$. We must still show that ψ is well defined, i.e. if $\Sigma\, d_i r_i = 0$, then $\psi(\Sigma\, d_i r_i) = 0$. $\Sigma\, d_i r_i = 0$ if and only if $(d_1,\ldots,d_n)$ is in $\ker\beta$, the kernel of β. If $(e_1,\ldots,e_n) \in (\ker\beta)_x$, then $\psi(\Sigma\, e_i r_i) = \psi(0) = 0$. But $f_1 = \alpha(1,0,\ldots,0),\ldots,f_m = \alpha(0,\ldots,0,1)$ generate $\ker\beta$ at all y near x by the exactness of (3.5). Hence for y sufficiently near x, $\psi(\beta(d_1,\ldots,d_n)) = 0$ since $(d_1,\ldots,d_n)$ is an ${_V\mathcal{O}_y}$-combination of the $(f_i)_y$. ∎

PROPOSITION 3.9. *If $\mathcal{R}$ and $\mathcal{S}$ are coherent analytic sheaves, then $\mathcal{H}om(\mathcal{R},\mathcal{S})$ is a coherent analytic sheaf.*

Proof: From the coherence of $\mathcal{R}$, we have locally an exact sequence

$${_V\mathcal{O}^m} \to {_V\mathcal{O}^n} \to \mathcal{R} \to 0\,.$$

By Proposition 3.8, which says what the stalks are,

$$0 \to \mathcal{H}om(\mathcal{R},\mathcal{S}) \to \mathcal{H}om({_V\mathcal{O}^n},\mathcal{S}) \to \mathcal{H}om({_V\mathcal{O}^m},\mathcal{S})$$

is exact. $\mathcal{H}om({}_V\mathcal{O}^n, \mathcal{S}) = \mathcal{S}^n$ and hence is coherent. Then $\mathcal{H}om(\mathcal{R}, \mathcal{S})$ is the kernel of a sheaf homomorphism between coherent sheaves. The image of such a sheaf homomorphism is coherent by Proposition IV. B.12 of G & R. Hence the kernel is coherent by Proposition VI. B.13 of G & R. ∎

Let us now return to the proof of Theorem 3.7. There is a natural inclusion ${}_V\mathcal{O} \subset \mathcal{H}om(\mathcal{I},\mathcal{I})$ given by $f(g) = fg$. This inclusion is a sheaf map of ${}_V\mathcal{O}$-modules. There is also a natural inclusion $\mathcal{H}om(\mathcal{I}, \mathcal{I}) \subset \tilde{\mathcal{O}}$ given as follows.

Let $\alpha \in \mathcal{H}om(\mathcal{I}, \mathcal{I})_x$ and let $f \in \mathcal{I}_x$ not be a zero divisor. Let $w = \frac{\alpha(f)}{f}$, an element of the total quotient ring. If g is also not a zero divisor, then $\alpha(fg) = \alpha(f)g = f\alpha(g)$ and hence g defines the same w. In the total quotient ring, if $h \in \mathcal{I}_x$, $wh = \frac{\alpha(f)h}{f} = \frac{\alpha(h)f}{f} = \alpha(h) \in \mathcal{I}_x$. Hence $w\mathcal{I}_x \subset \mathcal{I}_x$. Look at the ${}_V\mathcal{O}_x$-module M in the total quotient ring generated by ${}_V\mathcal{O}_x$, $w\ ({}_V\mathcal{O}_x)$, $w^2\,({}_V\mathcal{O}_x), \ldots$. Since $u \in \mathcal{I}_x$ is not a zero divisor and $w\mathcal{I}_x \subset \mathcal{I}_x$, multiplication by u maps M to a submodule of $\mathcal{I}_x$. Since ${}_V\mathcal{O}_x$ is Neotherian, M is finitely generated. Hence for sufficiently large n, w^n can be expressed in terms of ${}_V\mathcal{O}_x$, $w({}_V\mathcal{O}_x), \ldots, w^{n-1}({}_V\mathcal{O}_x)$, i.e. w is integral over ${}_V\mathcal{O}_x$. Hence by Theorem 3.6, $w \in \tilde{\mathcal{O}}_x$. If α is multiplication by a holomorphic function g, $w = g$. Hence we have the inclusions ${}_V\mathcal{O} \subset \mathrm{Hom}\,(\mathcal{I}, \mathcal{I}) \subset \tilde{\mathcal{O}}$.

Let $\mathcal{O}$ be the structure sheaf for some polydisc $\Delta \subset \mathbf{C}^n$ in which V is locally embedded. From II.B.20 of G & R, the Nullstellensatz for $\mathcal{O}_x$, we wish to verify the Nullstellensatz for ${}_V\mathcal{O}_x$. If F is an ideal in ${}_V\mathcal{O}_x$, we must show that id loc F = rad F. Let $\tilde{f}_1, \ldots, \tilde{f}_s$ generate F. Let $g_1, \ldots, g_t$ generate ${}_V\mathcal{I}_x$, as an ambient $\mathcal{O}$-module. Extend $\tilde{f}_1, \ldots \tilde{f}_s$ to germs $f_1, \ldots, f_s$ of functions in Δ. Suppose $\tilde{h} \in {}_V\mathcal{O}_x$ vanishes on the locus of F. Let h be an extension of $\tilde{h}$ to Δ. h vanishes on the locus of F, which is the locus of $f_1, \ldots, f_s, g_1, \ldots, g_t$. Hence by the ambient Nullstellensatz, for some k, $h^k = \Sigma\, r_i g_i + \Sigma\, r_j f_j$ for $r \in \mathcal{O}_x$. Now restrict back to V. $\tilde{h}^k = \Sigma\, r_j \tilde{f}_j$.

We next show that $N = \{x \mid {}_V\mathcal{O}_x \neq \mathcal{H}om(\mathcal{I}, \mathcal{I})_x\}$. If x is a normal point, then $\mathcal{O}_x = \tilde{\mathcal{O}}_x$ and hence, via the inclusions, ${}_V\mathcal{O}_x = \mathcal{H}om(\mathcal{I}, \mathcal{I})_x$. If x is not a normal point, ${}_V\mathcal{O}_x \neq \tilde{\mathcal{O}}_x$. $\mathcal{I}$ = id loc u. Hence there is a k such that $\mathcal{I}_x^k \subset \text{id } u_x$. Hence $\mathcal{I}_x^k \tilde{\mathcal{O}}_x \subset u_x \tilde{\mathcal{O}}_x \subset {}_V\mathcal{O}_x$ for all sufficiently large k. We may choose the smallest k such that $\mathcal{I}_x^k \tilde{\mathcal{O}}_x \subset {}_V\mathcal{O}_x$. $k \neq 0$ since $\tilde{\mathcal{O}}_x \not\subset {}_V\mathcal{O}_x$.

Let $w \in \mathcal{I}_x^{k-1} \tilde{\mathcal{O}}_x$, $w \notin {}_V\mathcal{O}_x$. Then $w \in \tilde{\mathcal{O}}_x$ and $w\mathcal{I}_x \subset {}_V\mathcal{O}_x$. We claim that $w: f \to wf$ is an element of $\mathcal{H}om(\mathcal{I}, \mathcal{I})_x$ which does not correspond to an element of ${}_V\mathcal{O}_x$. Since $w \notin {}_V\mathcal{O}_x$, $w: f \to wf$ does not correspond to an element of ${}_V\mathcal{O}_x$. We must still show that $w\mathcal{I}_x \subset \mathcal{I}_x$. w is locally bounded. $f \in \mathcal{I}_x$ implies that $f(z) \to 0$ as $z \to S$. Hence $wf(z) \to 0$ as $z \to S$. $wf \in {}_V\mathcal{O}_x$. Hence $wf \in \mathcal{I}_x$.

Thus $N = \{x \mid \mathcal{H}om(\mathcal{I}, \mathcal{I})/{}_V\mathcal{O} \neq 0\}$. However the support of any coherent analytic sheaf $\mathcal{R}$ is an analytic subvariety. Namely, if

$${}_V\mathcal{O}^m \overset{\phi}{\to} {}_V\mathcal{O}^n \to \mathcal{R} \to 0$$

is a local resolution of $\mathcal{R}$, $\mathcal{R}$ is non-trivial exactly where ϕ is not onto. This is a problem in linear algebra over the ring ${}_V\mathcal{O}_x$, where the units are non-vanishing functions. Hence ϕ is not onto precisely where the appropriate determinants vanish. This defines a subvariety. ▮

COROLLARY 3.10. *The set of non-normal points in a pure dimensional analytic space is open.*

LEMMA 3.11. $\tilde{\mathcal{O}}_x$ *is a finitely generated module over* ${}_V\mathcal{O}_x$ *if* V *is pure dimensional near* x.

Proof: Let u be a universal denominator near x. $u: h \to uh$ establishes a ${}_V\mathcal{O}_x$-module isomorphism of $\tilde{\mathcal{O}}_x$ with a submodule of the Neotherian ring ${}_V\mathcal{O}_x$. ▮

Proof of Theorem 3.5. By Theorem 3.4, any two normalizations of an analytic space agree. Thus Theorem 3.5 is purely local and we need only normalize V in a neighborhood of each point. The normalizations will then patch together in a unique manner.

Let π: $M \to V$ be a resolution of V such that except for a discrete set of points $P = \{p_i\}$ in V, $\pi^{-1}(p_i)$ consists of a finite number of points. Such a resolution was constructed in the proof of Theorem 2.1. π: $M - \pi^{-1}(P) \to V-P$ is a normalization of $V-P$. It thus suffices to find normalizations for each irreducible component of V near each p_i, for if $Y_1, \ldots, Y_t$ normalize the components, $Y = \cup Y_i$, a disjoint union, will normalize the subvariety.

So we are reduced to the case where π: $M \to V$ is a resolution of V, V is irreducible at p and $\pi^{-1}(p)$ is the only fibre which is not a finite number of points. Since R, the regular points of V, is connected near p, $\pi^{-1}(p)$ is connected by Lemma 3.2. π is biholomorphic off R. Let $f_1, \ldots, f_t$ be weakly holomorphic functions near p which generate $\tilde{\mathcal{O}}_p$ over $\mathcal{O}$. Let U be a neighborhood of p such that all the f_i are defined on $U \cap R$. The $\pi^* f_i$ are bounded functions on $\pi^{-1}(U \cap R) \subset \pi^{-1}(U)$ which are holomorphic except on a nowhere dense subvariety. Hence $\pi^* f_i$ are holomorphic functions, $\{F_i\}$, on $\pi^{-1}(U)$. We may assume that U is a subvariety of a polydisc and hence holomorphically convex. $\pi^{-1}(U)$ is then holomorphically convex. By Theorem VII. B. 9 of G & R there is a proper map ρ: $\pi^{-1}(U) \to S$ where S is a Stein space such that $\Gamma(S, \mathcal{O}) \approx \Gamma(\pi^{-1}(U), \mathcal{O})$. The underlying topological space of S is obtained by identifying points in $\pi^{-1}(U)$ which are not separated by holomorphic functions. Since $\Gamma(U, {}_V\mathcal{O})$ separates the points of U, ρ is one-to-one on $\pi^{-1}(R)$. Also, since π: $\pi^{-1}(U) \to U$ is given by an n-tuple of holomorphic functions, there is a unique proper holomorphic map ϕ: $S \to U$ such that $\pi = \phi \circ \rho$. $\pi^{-1} \circ \phi = \rho^{-1}$ on $\phi^{-1}(R)$ so that ρ is biholomorphic on $\pi^{-1}(R)$. Also, ϕ is biholomorphic on $\phi^{-1}(R)$. ϕ has finite fibres since off p, π has finite fibres. $\pi^{-1}(p)$ is a connected compact analytic set, so holomorphic functions are constant on $\pi^{-1}(p)$. Hence $\rho(\pi^{-1}(p)) = \phi^{-1}(p) = \tilde{p}$ is a single point.

We claim that S is normal at $\tilde{p}$. This will show that S is normal in a neighborhood of $\tilde{p}$ by Corollary 3.10 and hence $\phi: S \to V$ is a normalization of V in a neighborhood of p. $\{F_i\}$ are holomorphic on $\pi^{-1}(U)$ and hence on S. Let h be weakly holomorphic near $\tilde{p}$. By Lemma 3.2, there is a corresponding weakly holomorphic function $\phi_*(h)$ on V obtained by restricting h to $\phi^{-1}(R)$. $\phi_*(h) = \Sigma\, g_i f_i$ since the f_i generate $\tilde{\mathcal{O}}_p$. But the F_i are holomorphic on S and $\phi^*(g_i)$ is holomorphic near $\tilde{p}$. Hence h is in fact holomorphic near $\tilde{p}$. ∎

A few remarks are in order.

The normalization Y of an analytic space V was obtained from a resolution M by using Theorem VII. B. 9 of G & R to collapse certain analytic subsets to points. Since a normal space is its own normalization, we have

THEOREM 3.12. *The singular points of a 2-dimensional normal analytic space are isolated.*

THEOREM 3.13. *Let* p *and* p′ *be normal 2-dimensional singular points of* V *and* V′. *Let* $\pi: M \to V$ *and* $\pi': M' \to V'$ *be resolutions. If* $\pi^{-1}(p)$ *and* $(\pi')^{-1}(p')$ *have biholomorphic neighborhoods, then* p *and* p′ *have biholomorphic neighborhoods.*

Proof: V and V′ are normal in suitable neighborhoods of p and p′ by Corollary 3.10. Restricting to these neighborhoods, V and V′ are obtained by applying Theorem VII. B.9 of G & R to any sufficiently small neighborhoods of $\pi^{-1}(p)$ and $(\pi')^{-1}(p')$. These neighborhoods may be chosen to be biholomorphic by hypothesis. Hence V and V′ are biholomorphic near p and p′. ∎

M in the following theorem can be replaced by any normal space.

THEOREM 3.14. *If* $\pi: M \to V$ *is a resolution of* V *and* $\pi': Y \to V$ *is the normalization of* V, *then there is a unique holomorphic map* $\phi: M \to Y$ *such that* $\pi = \pi' \circ \phi$.

Proof: If R is the set of regular points of V, π is biholomorphic on $\pi^{-1}(R)$ and π' is biholomorphic on $(\pi')^{-1}(R)$. Hence ϕ is unique if it exists and it is merely a matter of extending $(\pi')^{-1} \circ \pi$ to all of M.

Suppose $a \in M - \pi^{-1}(R)$. Let $p = \pi(a)$ and $\{a_1, \ldots, a_s\} = (\pi')^{-1}(p)$. In $\pi^{-1}(R)$, $(\pi')^{-1} \circ \pi$ maps points near a to points near $(\pi')^{-1}(p)$. There are disjoint neighborhoods $U_1, \ldots, U_s$ of $a_1, \ldots, a_s$ which are biholomorphic to subvarieties of bounded polydiscs $\Delta_1, \ldots, \Delta_s$ in various dimensional C^n. By taking the largest n and making the Δ_i disjoint, we can embed $\cup U_i$ in a bounded part of C^n. $(\pi')^{-1} \circ \pi$ is then given by bounded holomorphic functions. Since M is normal, $(\pi')^{-1} \circ \pi$ may be extended to a holomorphic map on all of M. ∎

Theorem 3.14 shows that resolutions cannot distinguish isolated singular points which have the same normalizations. Resolutions thus really study normal singular points and Theorem 3.13 says that we can classify normal singular points by their resolutions. Theorem 3.12 says that this is a local problem in dimension two since normal singular points are isolated.

CHAPTER IV

EXCEPTIONAL SETS

The object of this chapter is to characterize the embeddings of 1-dimensional analytic sets which occur as $\pi^{-1}(p) \subset M$, when $\pi: M \to V$ is a resolution of a normal 2-dimensional singularity p.

It will be useful to take a somewhat more general outlook. We must concern ourselves not only with resolutions but also with maps that "resolve" a regular point.

DEFINITION 4.1. *Let* M *and* V *be analytic spaces. A proper holomorphic map* $\pi: M \to V$ *is called a point modification at* $p \in V$ *if* $\pi: M - \pi^{-1}(p) \to V - p$ *is biholomorphic and* $M - \pi^{-1}(p)$ *is dense in* M.

Thus iterations of quadratic transformations are point modifications, as are resolutions of analytic spaces with just one singular point.

LEMMA 4.1. *Let* $\pi: M \to V$ *be a point modification at* $p \in V$ *with* M *a manifold and* V_p *irreducible. Then* $A = \pi^{-1}(p)$ *is connected.*

Proof: Suppose U_1 and U_2 were open sets in M disconnecting A. The regular points R of V near p are connected. But then U_1 and U_2 would disconnect $\pi^{-1}(R)$ near $\pi^{-1}(A)$ by Lemma 3.2, an impossibility.∎

LEMMA 4.2. *Let* $S = (a_{i,j})$ $1 \leq i, j \leq n$ *be a symmetric real matrix and let* Q *be a non-singular real* $n \times n$ *matrix. Then* S *is negative definite or negative semi-definite if and only if* Q^tSQ *is negative definite or negative semi-definite.*

Proof: S is negative definite if and only if $X^tSX > 0$ for any non-zero $n \times 1$ vector X. $Y^t(Q^tSQ)Y = (QY)^tS(QY)$. Q: $\mathbf{R}^n \to \mathbf{R}^n$ is an isomorphism, so $X = QY$ for some unique vector Y. ∎

In actual practice, Lemma 4.2 says that we can perform row and column operations on a matrix and not destroy negative definiteness, provided that we follow a row operation immediately by the corresponding column operation.

LEMMA 4.3. *Let* A *be a non-singular compact* 1*-dimensional analytic subspace of* M, *a* 2*-dimensional complex manifold. Let* $\overline{\pi\colon M' \to M}$ *be a quadratic transformation at a point* $q \in A$. *Let* $A' = \pi^{-1}(A-q)$. *Then* $A' \cdot A' = A \cdot A - 1$.

Proof: $A \cdot A$ equals the Chern class of the normal bundle N (Theorem 2.3). As demonstrated before the proof of Theorem 2.6, if $\mathcal{I}$ is the ideal sheaf of A, $\mathcal{I}/\mathcal{I}^2 \approx \mathcal{N}^*$. Let f be a meromorphic section of N*, with f holomorphic near q. $-c(N)$ equals the algebraic sum of the zeros and poles of f. For an appropriately small neighborhood U of q in M, we may represent f by $F \in \Gamma(U, \mathcal{I}/\mathcal{I}^2)$. Choose coordinates (x,y) so that $A = \{y = 0\}$. The quadratic transformation is $(x,y) = (\xi\zeta, \zeta) = (\zeta', \xi'\zeta')$. $A' = \{\xi' = 0\}$. Locally we may represent F by the ambient holomorphic function $y\,F(x)$. Under π, $y\,F(x)$ becomes $\xi'\zeta'F(\zeta') \in \Gamma(U', \mathcal{I}/\mathcal{I}^2)$. Since π is biholomorphic off $\pi^{-1}(q)$, $\xi'\zeta'F(\zeta')$ induces a section of the dual to the normal bundle of A' which has the same zeros and poles away from $\pi^{-1}(q)$ as does f. $\xi'\zeta'F(\zeta')$ has an additional zero at $\zeta' = 0$. Hence $A' \cdot A' = A \cdot A - 1$. ∎

Given a resolution of an isolated singularity p, by performing additional quadratic transformations, we can always obtain a resolution such that the $\{A_i\}$, $1 \le i \le n$, the irreducible components of $A = \pi^{-1}(p)$, are non-singular and intersect transversely.

THEOREM 4.4. *Let* π: M $\to$ V *be a point modification of the 2-dimensional analytic space* V *at* $p \in V$ *with* M *a manifold,* V_p *irreducible and* p *an isolated singularity. Let* $\{A_i\}$, the irreducible components of $A = \pi^{-1}(p)$, *be non-singular and intersect transversely. Then* $S = (a_{ij}) = (A_i \cdot A_j)$, *the intersection matrix, is negative definite.*

Proof: Take a function f on V which is holomorphic in some neighborhood of p such that $f(p) = 0$, f is not identically zero and p is not an isolated zero of f. Such an f can be found, for example, by taking an admissible representation $\rho : V \to \Delta \subset C^2$ and letting $f = \rho^*(z_1)$. $F = \pi^*(f)$ is then holomorphic in a neighborhood N of A, $F(A) = 0$ and F vanishes at some points arbitrarily close to A. Let $W = \{q \in N \mid F(q) = 0\}$. It will be convenient to assume that in some neighborhood of A, the only singularities of W are where two submanifolds intersect transversely. By Theorem 1.1 and Proposition 1.2, this will occur if we perform suitable iterated quadratic transformations π' at points of A. However, this creates a new $A' = (\pi')^{-1}(A)$ as the inverse image of p. Let $\{A'_k\}$ be the set of irreducible components of A'.

We now show that $(A'_k \cdot A'_\ell)$ is negative definite if and only if $S = (A_i \cdot A_j)$ is negative definite. If a quadratic transformation π' is performed at a point $q \in A_1$, and q lies in no other A_j, then $[(\pi')^{-1}(q)] \cdot [(\pi')^{-1}(q)] = -1$, $[(\pi')^{-1}(q)] \cdot [(\pi')^{-1}(A_j)] = 0$, $j \neq 1$, $[(\pi')^{-1}(q)] \cdot [\overline{(\pi')^{-1}(A_1 - q)}] = 1$ and

$$[\overline{(\pi')^{-1}(A_1 - q)}] \cdot [\overline{(\pi')^{-1}(A_1 - q)}] = A_1 \cdot A_1 - 1 \quad \text{by}$$

Lemma 4.3. Thus if S looks like

$$\begin{pmatrix} A_1 \cdot A_1 & \cdots \\ \cdot & \\ \cdot & \\ \cdot & \end{pmatrix}, \quad (A_k' \cdot A_\ell') = \begin{pmatrix} -1 & 1 & 0 & 0 \cdots \\ 1 & A_1 . A_1 - 1 \cdots & & \\ 0 & \vdots & & \\ \vdots & \vdots & & \end{pmatrix}$$

A transformation as in Lemma 4.2 brings $(A_k' \cdot A_\ell')$ to the form

$$\begin{pmatrix} -1 & 0 & 0 \dots \\ 0 & A_1 \cdot A_1 & \dots \\ \vdots & \vdots & \end{pmatrix},$$

which is negative definite if and only if S is. If q is where $A_1, \dots, A_t$ meet, then $(A_k' \cdot A_\ell')$ looks like

$$\begin{pmatrix} -1 & 1 & \dots & 1 & 0 \dots \\ 1 & A_1 \cdot A_1 & \dots & A_1 \cdot A_t - 1 & 0 \dots \\ \vdots & \vdots & & & \\ 1 & & & & \\ 0 & & & & \\ \vdots & & & & \end{pmatrix}$$ which, as before,

is negative definite if and only if S is.

Thus it suffices to prove that $(A_k' \cdot A_\ell')$ is negative definite. Let m_k be the order of the zero of $(\pi')^*F$ on A_k'. Let $S' = (m_k m_\ell A_k' \cdot A_\ell') = (\alpha'_{k\ell})$. It suffices to prove that S' is negative definite. By Theorem 2.6,

$$0 = ((\pi')^*F) \cdot A_\ell \geq \sum_{k=1}^{n'} m_k A_k \cdot A_\ell .$$ Hence

a) $\sum_{k=1}^{n'} \alpha'_{k\ell} \leq 0 .$ Also

b) $\alpha'_{k\ell} \geq 0$ if $k \neq \ell$.

Finally since W must contain irreducible components besides the A_k, by Theorem 2.6, for an A_ℓ' which meets one of these components,

c) $\sum_{k=1}^{n'} \alpha'_{k\ell} < 0 .$

a) and b) imply that S' is negative semi-definite as follows. If $\alpha'_{11} = 0$, a) and b) imply that $\alpha'_{k1} = 0$ for all k. We can thus disregard the first row and column. We then proceed by induction on n', assuming $\alpha'_{11} \neq 0$. Then $\alpha'_{11} < 0$. S' is negative definite if $n' = 1$. For general values of n', apply Lemma 4.2 with

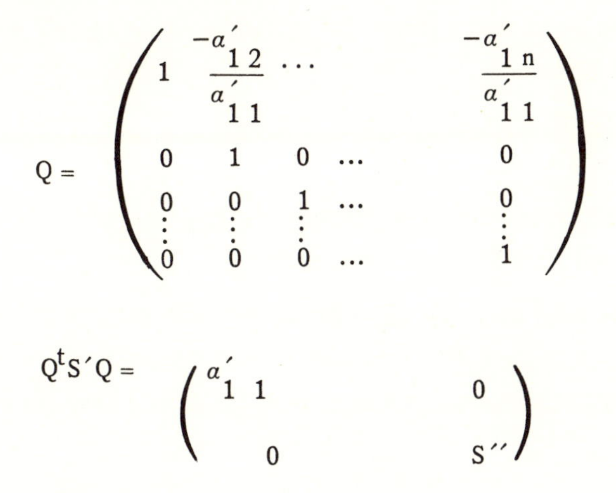

$$Q = \begin{pmatrix} 1 & \frac{-\alpha'_{12}}{\alpha'_{11}} & \cdots & & \frac{-\alpha'_{1n}}{\alpha'_{11}} \\ 0 & 1 & 0 & \cdots & 0 \\ 0 & 0 & 1 & \cdots & 0 \\ \vdots & \vdots & \vdots & & \vdots \\ 0 & 0 & 0 & \cdots & 1 \end{pmatrix} .$$

$$Q^t S' Q = \begin{pmatrix} \alpha'_{11} & 0 \\ 0 & S'' \end{pmatrix}$$

with S'' also having properties a) and b).

We prove that S' is definite using c) and the connectedness of A' (Lemma 4.1). Suppose

$$0 = \sum_{\ell,k} x_k x_\ell \alpha'_{k\ell} = \sum_k x_k^2 \alpha'_{kk} + 2 \sum_{k<\ell} x_k x_\ell \alpha'_{k\ell}$$

$$= \sum_\ell \left(\sum_k \alpha'_{k\ell}\right) x_\ell^2 - \sum_{k<\ell} \alpha'_{k\ell} (x_k - x_\ell)^2 .$$

For some ℓ, $\sum_k \alpha'_{k\ell} < 0$, so $x_\ell = 0$. A' is connected so that for some ℓ', $\alpha'_{\ell'\ell} > 0$. Then $x_{\ell'} = 0$. $A'_{\ell'}$ meets some $A'_{\ell''}$ so $x_{\ell''} = 0$. After n' steps, since A' is connected, $x_k = 0$ for all k. Hence S' is negative definite. ∎

To say that an analytic set A is nowhere discrete is the same as saying that $\dim A_q > 0$ for each point $q \in A$ since all 0-dimensional analytic spaces are discrete.

DEFINITION 4.2. *A nowhere discrete compact analytic subset* A *of an analytic space* G *is called exceptional (in* G) *if there exists an analytic space* Y *and a proper holomorphic map* $\Phi : G \to Y$ *such that* $\Phi(A)$ *is discrete,* $\Phi : G - A \to Y - \Phi(A)$ *is biholomorphic and such that for any open set* $U \subset Y$, *with* $V = \Phi^{-1}(U)$, $\Phi^* : \Gamma(U,\mathcal{O}) \to \Gamma V,\mathcal{O})$ *is an isomorphism.*

If A is exceptional in G, we shall sometimes say that A can be "blown down", or Φ blows down A.

If for some $p \in A$, $\dim A_p = \dim G_i$ on some pure dimensional component G_i of G, then $\Phi(G_i)$ is just a point, a trivial case in which we are not really interested. Thus in general, A will have positive codimension.

Φ^* is always an injection so Definition 4.2 only really requires that Φ^* be onto. On the other hand, analytic spaces can be locally embedded in some $\mathbf{C}^n$ and hence $\Gamma(U,\mathcal{O})$ will be large for small neighborhoods U of $\Phi(A)$. Hence in order for A to be exceptional, A must have many global ambient holomorphic functions. If A is exceptional, to find Y one must construct an analytic space with sufficient holomorphic functions for Φ^* to be surjective.

PROPOSITION 4.5. *If* $\pi : M \to V$ *is a point modification at* p, *a normal point of* V, *such that* $A = \pi^{-1}(p)$ *is nowhere discrete, then* A *is exceptional in* M.

Proof: We need only verify that for an open set $U \ni p$, $\pi^* : \Gamma(U,\mathcal{O}) \to \Gamma(\pi^{-1}(U),\mathcal{O})$ is surjective. A is compact. Hence f is bounded near $\pi^{-1}(p)$. Restrict f to $\pi^{-1}(U-p)$. π is biholomorphic so we can get a holomorphic function g on $U-p$ which is bounded near p. V is normal at p, so g extends to a holomorphic function g at p. $\pi^*(g)=f.$ ∎

PROPOSITION 4.6. *If in Definition 4.2* G *is normal, then* Y *is normal.*

Proof: Consider G_i, a global irreducible component of G. G is normal and hence locally irreducible so G_i is a connected component of G. If A is not of positive codimension in G_i, $\Phi(G_i)$ is a point. If $\Phi(G_i)$ is an isolated point of Y, then Y is normal at $\Phi(G_i)$. If $\Phi(G_i)$ is not an isolated point, since Φ is a proper map, $\Phi(G_i) \subset \Phi(G_j)$ for some $G_j \neq G_i$. Hence we may assume that A has positive codimension.

We must show that Y is normal for $p \in \Phi(A)$. If f is weakly holomorphic near p, then f is holomorphic at the regular points R of some neighborhood N of p and bounded in N. Then $\Phi^*(f)$ is holomorphic on $\Phi^{-1}(R)$ and bounded in $\Phi^{-1}(N)$. At any regular point q of G, $q \in A$, the Riemann removable singularity theorem extends $\Phi^*(f)$ through q. Since G is normal and $\Phi^*(f)$ is now bounded and holomorphic at all regular points, $\Phi^*(f)$ extends to be holomorphic on $\Phi^{-1}(N)$. Φ^* is surjective, so f is holomorphic on N.▌

LEMMA 4.7. *Let* A *be a nowhere discrete compact analytic subset of a 2-dimensional manifold* M *and* $\pi: M' \to M$ *a quadratic transformation at a point of* p *of* A. *Then* A *is exceptional in* M *if and only if* $\pi^{-1}(A)$ *is exceptional in* M′.

Proof: By Proposition 4.5, for an open set U in M, $\pi^*: \Gamma(U, \mathcal{O}) \to (\pi^{-1}(U), \mathcal{O})$ is an isomorphism. The lemma now follows easily.▌

We shall explicitly compute some examples, omitting the verification of many details. The main problem is to make Φ^* is surjective. Since Φ is biholomorphic off A and sheaf sections may be defined locally, it suffices to make $\Phi^*(\Gamma(\Phi(A), \mathcal{O})) = \Gamma(A, \mathcal{O})$, where these sections are actually defined in neighborhoods of $\Phi(A)$ and A.

First consider M(−2), as introduced in Section II, which we already know has A as an exceptional set. There are two coordinate patches, (u,v) and (u′,v′) with $u = \frac{1}{u'}$, $v = u'^2 v$. Let us determine $\Gamma(A,\mathcal{O})$. In (u,v), $f \in \Gamma(A,\mathcal{O})$ may be expanded in a power series,

$$f = \sum_{\substack{k \geq 0 \\ \ell \geq 0}} a_{k\ell} u^k v^\ell \quad ;$$

we shall not have to worry about convergence conditions. In (u′,v′), $f = \Sigma\, a_{k\ell} u'^{2\ell - k} v'^{\ell}$ so we must have $2\ell - k \geq 0$. The points (k,ℓ) representing permissible exponents are depicted below. They form a cone.

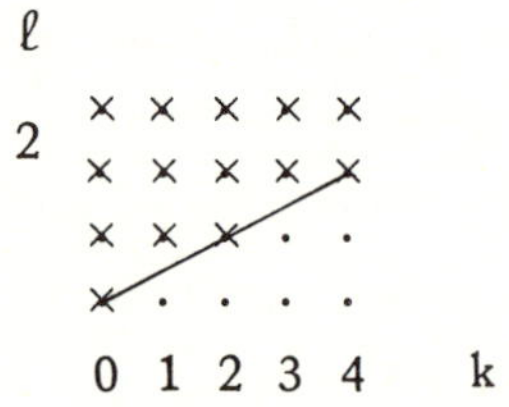

```
ℓ
    ×  ×  ×  ×  ×
2   ×  ×  ×  ×  ×
    ×  ×  ×  ·  ·
    ×  ·  ·  ·  ·
    0  1  2  3  4     k
```

(0,1), (1,1) and (2,1) generate the cone, so v, uv and u^2v generate $\Gamma(A,\mathcal{O})$. Thus the appropriate map $\Phi : M \to V$ is given by $\Phi(u,v) = (x,y,z) = (u,uv,u^2v)$. $V = \{y^2 = xz\}$. One can now verify all the required conditions for Φ to represent A as exceptional.

The same computations work for M(-r). V is embedded in $\mathbb{C}^{r+1}$ by $v, uv, \ldots, u^r v$.

Looking at M(−4), we see that $\Phi(u,v) = (v,uv,u^3v,u^4v)$ does not blow down A since u^2v is not holomorphic on V, the image space. However Φ is proper and biholomorphic off A. V is not normal at $0 = (0,0,0,0)$ since u^2v is only weakly holomorphic. Thus 0 is an isolated singular point which is not a normal point. Theorem 3.1 insures, of course, that this cannot happen in $\mathbb{C}^3$.

Next consider M(−2,−2). (u,v), (u′,v′) and (u″,v″) are the coordinate patches. Calculating as with M(−2), $f \in \Gamma(A,\mathcal{O})$ must be represented as $f \in \Sigma\, a_{k\ell} u'^k v'^\ell$ with $2\ell - k \geq 0$ and $2k - \ell \geq 0$. As before, points (k, ℓ) representing exponents with non-zero coefficients form a cone.

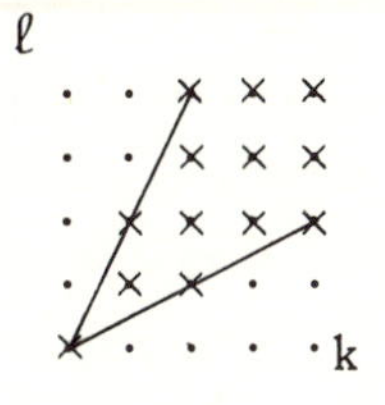

(1,1), (2,1) and (1,2) generate the cone, so $u'v'$, u'^2v' and $u'v'^2$ generate $\Gamma(A, \mathcal{O})$. Thus the appropriate map is
$\Phi(u', v') = (x, y, z) = (u'v', u'^2v', u'v'^2)$. $V = \{x^3 = yz\}$.

The reader may amuse himself by blowing down $M(k_1, \ldots, k_s)$ in the above manner.

Let us finish with a different kind of example. Let A be an elliptic curve, q a point on A and M the total space of the line bundle corresponding to the divisor q. Then sections of M are holomorphic functions on A with zeros of order at least 1 at q. Let m be the ideal sheaf of A in M. We must represent $\Gamma(A, \mathcal{O})$. Given $f \in \Gamma(A, \mathcal{O})$ expand f in a power series as follows. f is constant on A and thus determines $f_o \in \Gamma(A, C)$. $f - f_o$ determines $f_1 \in \Gamma(A, m/m^2)$. On the other hand, the line bundle corresponding to m/m^2 is dual to M and thus has sections which may be identified with meromorphic functions on M which have poles of order at most one at q. Thus given $f_1 \in \Gamma(A, m/m^2)$, we get a holomorphic function on M as follows. Let s and t be local coordinates for M with s a coordinate for A and t a fibre coordinate. Let s and t also be the coordinates for M*. (s,t) will, of course, transform differently in M and M*. Locally f_1 is given $t = f_1(s)$. Let $f_1(s,t) = t\, f_1(s)$. As verified in the proof of Lemma 4.11 to follow, $f_1(s,t)$ is in fact a well defined holomorphic function on M. Moreover $f_1(s,t)$ and f_1 determine the same section in $\Gamma(A, m/m^2)$). Next $f - f_o - f_1(s,t)$ determines

$f_2 \in \Gamma(A, m^2/m^3) = \Gamma(A, \mathfrak{M}^* \otimes \mathfrak{M}^*)$. $f_2(s,t) = t^2 f_2(s)$ is again a well-defined holomorphic function on M. We may continue the argument. Hence $\Gamma(A, m/m^2)$, $\Gamma(A, m^2/m^3), \ldots$ generate $\Gamma(A, \mathcal{O})$ (formally at least). We need a finite set of generators.

The bundles M,M* and their tensor powers are product bundles over $A - q$. Holomorphic sections are determined by their restrictions to $A - q$ so it suffices to represent sections only over $A - q$. $\mathbb{C}$ is the universal covering space for A. Let (z,t) be coordinates for M over $A - q$, with z the coordinate for $\mathbb{C}$. Let $z = 0$ project onto q. If (z,t') are coordinates for M near q, the transition functions are $t = t'z$, $z = z$. Sections of M* and its tensor powers over $A - q$ are conveniently represented on $\mathbb{C}$ by elliptic functions (see [Hℓ, Chap. 13]). These are doubly periodic meromorphic functions on $\mathbb{C}$, with poles only at the origin and its images under deck transformations, i.e. the lattice points. Let us recall the basic facts about elliptic functions with poles only at the lattice points. There are no non-constant ones with only first order poles. The Weierstrass $\wp$-function is non-constant with a pole of order 2. $\wp'$ is non-constant with a pole of order 3.

$$(\wp'(z))^2 - 4[\wp(z)]^3 + g_2\wp(z) + g_3 = 0 \tag{4.1}$$

where g_2 and g_3 are complex constants depending on the lattice (and hence on A).

Since $\wp$ has a pole of order 2, $\wp'$ a pole of order 3 and elliptic functions cannot have a single pole of order 1, linear combinations of products of $\wp$ and $\wp'$ give all elliptic functions with poles only at the lattice points. Thus, in (z,t) coordinates, $t \cdot 1 \in \Gamma(A, m/m^2)$, $t^2 \cdot \wp(z) \in \Gamma(A, m^2/m^3)$ and $t^3 \cdot \wp'(z) \in \Gamma(A, m^3/m^4)$ generate, via tensor product of sections and addition, $\Gamma(A, m/m^2)$, $\Gamma(A, m^2/m^2), \ldots$

and hence generate $\Gamma(A,\mathcal{O})$, at least formally. A may then be blown down by $\Phi(z,t) = (x_1,x_2,x_3) = (t, t^2\wp(z), t^3\wp'(z))$. We may check, of course, that under $t = t'z$, the transition rules for M, t, $t^2\wp(z)$ and $t^3\wp'(z)$ are holomorphic on M. We use (4.1) to calculate the defining equation for $V = \Phi(M)$.

$$[\wp'(z)]^2 t^6 - 4[\wp(z)]^3 t^6 + g_2\, \wp(z) t^6 + g_3 t^6 = 0$$

Hence $V = \{x_3{}^2 - 4x_2{}^3 + g_2 x_2 x_1{}^4 + g_3 x_1{}^6 = 0\}$. Thus when $g_2 = 0$, we get one of our examples from section II. To complete the verification of this example, one must show that Φ is biholomorphic off A and that 0 is an isolated singularity of V and hence normal. Proposition 4.5 then insures that our formal computations are correct.

DEFINITION 4.3. *A nowhere discrete compact analytic subset* A *in an analytic space* G *is the maximal compact analytic set in* G *if* $A \supset A'$ *for any nowhere discrete compact analytic set* A' *in* G.

THEOREM 4.8. *Let* G *be an analytic space and* A *a compact, nowhere discrete analytic subset.* A *is exceptional if and only if there exists a neighborhood* U *of* A *such that the closure of* U *in* G *is compact,* U *is strictly Levi pseudoconvex and* A *is the maximal compact analytic subset of* U. *Also,* A *is exceptional if and only if* A *has arbitrarily small strictly Levi pseudoconvex neighborhoods.*

Proof: Suppose that $\Phi : G \to Y$ exhibits A as an exceptional set. Let $\Phi(A) = \{p_1, \ldots, p_\ell\}$. Y is an analytic space so it can be embedded in some C^r near each p_i, with $p_i = (0, \ldots 0)$. Given $\varepsilon > 0$, around each p_i choose a spherical neighborhood $N_i(\varepsilon) = \{z \in Y \mid z_1\bar{z}_r < \varepsilon\}$. For ε small enough so that $p_j \notin \bar{N}_i(\varepsilon)$ for $j \neq i$ and $\bar{N}_i(\varepsilon)$ compact in Y, $U_\varepsilon = \bigcup_{i=1}^{\ell} \Phi^{-1}(N_i(\varepsilon))$ is the required neighborhood. U is strictly Levi pseudoconvex (Definition IX.B.8 of G & R) since $z_1\bar{z}_1 + \ldots + z_r\bar{z}_r$ is strictly plurisubharmonic. $\bar{U}_\varepsilon = \cup\, \Phi^{-1}(\overline{N_i(\varepsilon)})$ and hence is compact.

Since holomorphic functions are constant on connected, compact analytic sets and $\Phi|U_\varepsilon$ is given by holomorphic functions, $\cup \Phi^{-1}(p_i) = A$ must contain all compact nowhere discrete analytic sets. As $\varepsilon \to 0$, U_ε becomes an arbitrarily small neighborhood of A.

For the first converse, suppose that U exists with A the maximal compact analytic subset. By Theorem IX.C.4 of G & R, there is a Stein space X and a proper holomorphic map $\pi : U \to X$ such that (i) $\pi^* : \Gamma(X, \mathcal{O}) \to \Gamma(U, \mathcal{O})$ is an isomorphism and (ii) there are finitely many points $x_1, \ldots x_t$ in X such that $\pi^{-1}(x_t)$ is a compact subvariety of U of positive dimension and $\pi : U - \cup \pi^{-1}(x_i) \to X - \{x_1, \ldots, x_r\}$ is biholomorphic. Topologically, π is obtained by identifying points in U which cannot be separated by holomorphic functions. Thus any compact analytic set of positive dimension is included $\cup \pi^{-1}(x_t)$. Since A is maximal, $\cup \pi^{-1}(x_t)$ is the union of A and a discrete set B. As in the first part of the proof, we may choose small neighborhoods $N_i(\varepsilon)$ of the $\{x_i\}$. For small ε, $\cup \pi^{-1}(N_i(\varepsilon))$ will decompose into open subsets which are respectively neighborhoods of A and B. Thus by restricting π to the neighborhood of A, we may assume that π is biholomorphic off A in some neighborhood of A. We can then patch π with the identity map to get $\Phi : G \to Y$ such that Φ is proper and holomorphic, $\Phi(A)$ is discrete and $\Phi : G - A \to Y - \Phi(A)$ is biholomorphic. The trouble is that Φ^*, the induced map on holomorphic functions, may not yet be onto. Since holomorphic functions may be defined locally and Φ is biholomorphic off A, we need that $\Phi^* : \Phi(\cup x_i, \mathcal{O}) \to \Gamma(A, \mathcal{O})$ is surjective. We will show that this occurs for all sufficiently small $\varepsilon . U_\varepsilon = \cup \pi^{-1}(N_i(\varepsilon))$ is strictly Levi pseudo-convex. Since π is proper, for sufficiently small ε, U_ε will decompose into disjoint neighborhoods of the connected components of A. Apply Theorem IX.C.4 of G & R again. U_ε has more global holomorphic functions than has U. Thus $\pi_\varepsilon : U_\varepsilon \to X_\varepsilon$ satisfies $\pi = \phi_\varepsilon \circ \pi_\varepsilon$ for a unique holomorphic map $\phi_\varepsilon : X_\varepsilon \to X$ with ϕ_ε a homeomorphism of

$X_\varepsilon - \pi_\varepsilon(A)$ onto its image. Also, under π_ε, different components of A are separated by holomorphic functions which are constant on each component of U_ε. Taking sufficiently small ε, say $\varepsilon \leq \eta$, and working on each component separately, we may assume that A is connected and $\pi_\varepsilon(A) = x$, one point. As before, for $\varepsilon < \eta$, $\pi_\eta = \chi_\varepsilon \circ \pi_\varepsilon$ with χ_ε a homeomorphism on $X_\varepsilon - \pi_\varepsilon(A)$. Since $\pi_\varepsilon(A)$ is just one point, χ_ε is in fact a homeomorphism of all of X_ε onto its image. Also, since π_η and π_ε are biholomorphic except off the inverse image of a discrete set, χ_ε is biholomorphic on some deleted neighborhood of $\pi_\varepsilon(A)$. Let $\mathcal{O}(\varepsilon)$ be the germs at $\pi_\varepsilon(A)$ of holomorphic functions on X_ε. $\chi_\varepsilon^*: \mathcal{O}(\eta) \to \mathcal{O}(\varepsilon)$ is an injection. Similarly, for $\varepsilon_1 > \varepsilon_2 > \varepsilon_3 > \dots$, there are natural inclusions $\mathcal{O}(\varepsilon_1) \subset \mathcal{O}(\varepsilon_2) \subset \dots$. Every $f \in \Gamma(A, \mathcal{O})$ is represented by a function in some sufficiently small neighborhood U_ε of A. Thus f is represented in $\mathcal{O}(\varepsilon)$ for sufficiently small ε. Thus it suffices to show that for $\varepsilon_i \to 0$, $\mathcal{O}(\varepsilon_1) \subset \mathcal{O}(\varepsilon_2) \subset \mathcal{O}(\varepsilon_3) \subset \dots$ is eventually stationary. By restricting functions in $\mathcal{O}(\varepsilon_i)$ to the nearby regular points of x_η, we get weakly holomorphic functions on X_η, i.e. $\mathcal{O}(\varepsilon_i) \subset \tilde{\mathcal{O}}_x$, the germs at $\pi_\eta(A) = x$ of weakly holomorphic functions on X_η. It thus suffices to show that $\tilde{\mathcal{O}}_x$ is finitely generated over $\mathcal{O}(\eta)$, a Noetherian ring. Decompose X_η near x into irreducible components $V_1, \dots, V_r$ by decomposing the regular points into connected components. We may write $\tilde{\mathcal{O}}_x \approx \tilde{\mathcal{O}}_1 \oplus \dots \oplus \tilde{\mathcal{O}}_r$ where $\tilde{\mathcal{O}}_j$ are those weakly holomorphic functions where are identically 0 on the regular points not in V_j. By Lemma 3.11, $\tilde{\mathcal{O}}_j$ is finitely generately over ${}_{V_j}\mathcal{O}_x$. The restriction map $\mathcal{O}(\eta) \to {}_{V_j}\mathcal{O}_x$ is surjective, so $\tilde{\mathcal{O}}_j$ is finitely generated over $\mathcal{O}(\eta)$. Hence $\tilde{\mathcal{O}}_x$ is finitely generated over $\mathcal{O}(\eta)$. Hence A is exceptional.

To prove the second converse, we show that for sufficiently small neighborhoods, A is the maximal compact analytic subset. Given U, strictly Levi pseudoconvex, apply as usual Theorem IX.C.4 of G & R. $A \subset U\pi^{-1}(x_i)$ and $\tilde{A}$, the non-discrete part of the $U\pi^{-1}(x_i)$, is the maximal compact subset of U. We need that $A = \tilde{A}$. Suppose that some irreducible component C of $\tilde{A}$ (obtained by taking the closure of a connected component of the regular points of $\tilde{A}$) is not contained in A. Choose a smaller strictly Levi pseudoconvex neighborhood U' of A such that $C \not\subset U'$. Let $\tilde{A}'$ be the maximal compact subset of U'. Then $\tilde{A}' \cap C$ is a proper subvariety of C and hence has positive codimension in C. Thus after a finite number of steps we get that each irreducible component C of $\tilde{A}$ with $C \not\subset A$ has dimension 0, which is impossible. Thus for sufficiently small neighborhoods, $A = \tilde{A}$ and thus A is exceptional.∎

THEOREM 4.9. *Let* A *be a compact purely* 1*-dimensional analytic subset of a* 2*-dimensional manifold* M. *Perform quadratic transformations* $\pi: M' \to M$ *until the* $\{A'_j\}$, *the irreducible components of* $A' = \pi^{-1}(A)$, *are non-singular and intersect transversely. Then* A *is exceptional if and only if* $S = (A'_i \cdot A'_j) = (\alpha'_{ij}), 1 \leq i, j \leq n$, *is a negative definite matrix.*

Proof: By Lemma 4.7, A is exceptional if and only if A' is exceptional. If A (and hence A') is exceptional, let $\Phi : M' \to V$ blow down A'. By Proposition 4.6, since M' is normal, V is normal. By Lemma 4.1 each connected component of A' is mapped to a different point under Φ. Then by Theorem 4.4 the intersection matrix for each connected component of A' is negative definite. Hence S is negative definite.

In proving the converse, it suffices to prove the theorem for each connected component of A'. As shown in the proof of Theorem 4.4, quadratic transformations do not change the negative definiteness of S. Lemma 4.7 shows that quadratic transformations do not change the exceptionality of A. Hence we may assume that no three A'_i meet at the same point and $A'_i \cap A'_j$, $i \neq j$, contains at most one point.

We proceed via a series of lemmas.

LEMMA 4.10. *If* $S = (\alpha_{ij})$, $1 \leq i, j \leq n$, *is negative definite and* $\alpha_{ij} \geq 0$ *if* $i \neq j$, *then there exist natural numbers* $r_1, \ldots, r_n$ *such that*

$$\sum_{i=1}^{n} r_i \alpha_{ij} < 0, \; 1 \leq j \leq n.$$

Proof: We show this by induction on n. It is trivally true if $n = 1$. Assume Lemma 4.10 for $n-1$. Multiplying S by $\frac{-1}{\alpha_{nn}}$, we may assume without loss of generality that $\alpha_{nn} = -1$. By Lemma 4.2, the matrix $S' = (\beta_{ij})$, where $\beta_{ij} = \alpha_{ij} + \alpha_{in} \cdot \alpha_{nj}$ for $i,j \leq n-1$, $\beta_{ij} = 0$, $i = n$, or $j = n$ but $i \neq j$ and $\beta_{nn} = -1$, is negative definite. S' satisfies the hypotheses of the lemma. Hence by the induction hypothesis, there exist positive integers $R_1, \ldots, R_{n-1}$ such that

$$(4.2) \quad \sum_{i=1}^{n-1} R_i \beta_{ij} = \sum_{i=1}^{n-1} R_i \alpha_{ij} + \alpha_{nj} \sum_{i=1}^{n-1} R_i \alpha_{in} < 0 \text{ for } 1 \leq j \leq n-1.$$

Let R_n be the positive real number $\varepsilon + \sum_{i=1}^{n-1} R_i \alpha_{in}$, with ε a small positive number to be chosen later. Then

$$\sum_{i=1}^{n} R_i \alpha_{ij} = \sum_{i=1}^{n-1} R_i \alpha_{ij} + \alpha_{nj} \sum_{i=1}^{n-1} R_i \alpha_{in} + \varepsilon \alpha_{nj}, 1 \leq j \leq n.$$

When $j = n$, since $\alpha_{nn} = -1$, $\sum_{i=1}^{n} R_i \alpha_{ij} = -\varepsilon < 0$.

By (4.2), $\sum_{i=1}^{n} R_i \alpha_{ij} < 0$ for all $j = 1, \ldots, n$ if ε is chosen small enough. To get positive integers $r_1, \ldots, r_n$, it suffices to approximate the ratios $R_1 : R_2 : \ldots : R_n$ by sufficiently closely by positive integers $r_1 : r_2 : \ldots : r_n$. ∎

Let p_i be the ideal sheaf on M of germs of functions vanishing on A_i'. Let $m = p_1^{r_1} \ldots p_n^{r_n}$. m is then a locally free sheaf of rank 1 on M and thus corresponds naturally to a line bundle V over M (Theorem VIII.C.6 of G & R) with m the sheaf of germs of sections of V. Let V* be the dual bundle to V. m*, the sheaf of germs of holomorphic sections of V* corresponds to germs of meromorphic functions on M with poles of order at most $r_1 \ldots r_n$ on $A_1' \ldots A_n'$ respectively.

We wish to calculate $c(V_i^*)$, the Chern class of the restriction V_i^* of V* to A_i'. Sections of V_i are given by m/mp_i. m/mp_i is the subsheaf of $p_i^{r_i}/p_i^{r_i+1}$ given by germs of sections which vanish to the r_j^{th}-order at $A_j \cap A_i$, $j \neq i$. $p_i^{r_i}/p_i^{r_i+1}$ corresponds to the r_i^{th} tensor power of the dual to the normal bundle of the embedding of A_i' in M (as shown before Theorem 2.6). Hence

$$c(V_i^*) = -c(V_i) = -[-\sum_{j=1}^{n} r_j A_j \cdot A_i] = \sum r_i \alpha_{j\,i} < 0 .$$

Let G be the restriction of V* to A′. Let G_i be the restriction of G to A_i'. From now on, we shall omit the primes.

LEMMA 4.11. *Let* G *be a line bundle over the compact* 1*-dimensional analytic space* A. *Suppose that* $\{A_i\}$*, the irreducible components of* A, *are non-singular and meet transversely. Suppose also that no three* A_j *meet at a point and* $A_i \cap A_j$ *is at most one point for* $i \neq j$. *If* $c(G_i) < 0$ *for all* i, *then* A *is exceptional in* G.

Proof: We will use Theorem 4.8 but nonetheless it is still necessary to create some holomorphic functions on G. Let $\underset{\nu}{\otimes} G$ denote $G \otimes \ldots \otimes G$ with ν factors. Let $\underset{\nu}{\otimes} \mathcal{G}$ be the corresponding sheaf of sections. Suppose $f \in \Gamma(A, \underset{\nu}{\otimes} \mathcal{G}^*)$. Let (s,t) be a coordinate patch on G with t the fibre coordinate and s the base coordinate on A. Then we may also use (s,t) as local coordinates for $\underset{\nu}{\otimes} G^*$. Locally f is given by $t = f(s)$. $F(s,t) = t^{\nu} f(s)$ is in fact a well defined holomorphic function on G. We must check what happens under a change of coordinates for G. Suppose then that $(s,t) = (g(s'), h(s')t')$ where $g(s') = s$ is the transition rule for A and $h(s') \neq 0$ gives the change for the fibre coordinate in G. Then in $\underset{\nu}{\otimes} G^*$, if T' is the new fibre coordinate, $(s,t) = (g(s'), h(s')^{-\nu} T')$. $F(s,t) = t^{\nu} f(s) = t'^{\nu} h(s')^{\nu} f(g(s'))$. On the other hand, in the (s',T') system, $t = f(s)$ becomes $T' = h(s')^{\nu} f(g(s'))$. So identifying t' with T' gives the same holomorphic function $F(s,t')$. Thus we create holomorphic functions by finding sections of $\underset{\nu}{\otimes} G^*$.

Since $c(G_i) < 0$, $c(\underset{\nu}{\otimes} G_i^*) \to \infty$ as $\nu \to \infty$. Thus for large ν, Riemann-Roch [Gu, p. 111] gives $\dim \Gamma(A_i, \underset{\nu}{\otimes} \mathcal{G}_i^*) = -\nu\, c(G_i) + 1 - g_i$, where g_i is the genus of A_i. Sections over A_i and A_j will join to form a holomorphic section over $A_i \cup A_j$ if they agree on the fibre above $A_i \cap A_j$. We shall find holomorphic functions $F_1, \ldots, F_\mu$ such that

$$U(\varepsilon) = \{ F_1 \overline{F}_1 + \ldots + F_\mu \overline{F}_\mu < \varepsilon \}$$

is a strictly Levi pseudoconvex neighborhood of A for all sufficiently small ε. Let us first show that $F_i \overline{F}_i$ is always plurisubharmonic. Locally embed G in $\Delta \subset \mathbf{C}^n$. F_i is the restriction to G of an ambient holomorphic function H. $H\overline{H}$ is plurisubharmonic since on any 1-dimensional subspace it is subharmonic. Thus to make $\Sigma F_i \overline{F}_i$ strictly plurisubharmonic near a point we only have to make some subtotal of $\Sigma F_i \overline{F}_i$ strictly plurisubharmonic. At each point $q \in A$, we shall find two (or at $A_i \cap A_j$, three) functions which map A into the origin and embed $N-A$ into $\mathbb{C}^2$ (or $\mathbb{C}^3$) for some neighborhood N of q. ν shall be chosen large, but independent of q.

The first case is where $q \in A_i$ is a regular point of A. Choose ν large enough so that there exist $f_1, f_2 \in \Gamma(A_i, \underset{\nu}{\otimes} \mathcal{G})$ such that f_1 and f_2 vanish at all $A_i \cap A_j$, $j \neq i$, (this imposes only a finite codimension condition, independent of ν, on $\Gamma(A_i, \underset{\nu}{\otimes} \mathcal{G}_i))$ f_1 vanishes to exactly first order at q and $f_2(q) \neq 0$. The last two conditions may be satisfied since, by Riemann-Roch, those sections vanishing to at least first and at least second order form subspaces of codimension 1 and 2 respectively. We can extend the sections f_1 and f_2 by 0 to get elements of $\Gamma(A, \underset{\nu}{\otimes} \mathcal{G}^*)$. We can choose local coordinates with $q = (0,0)$ so that $t = f_1(s) = s$. Then $t = f_2(s) = a + a_1 s + \ldots$ with $a \neq 0$. The corresponding functions F_1 and F_2 are $F_1(s,t) = st^{\nu}$ and $F_2(s,t) = (a + a_1 s + \ldots)t^{\nu}$. The map (F_1, F_2) has non-zero Jacobian for $t \neq 0$ and s small. Hence $F_1\overline{F}_1 + F_2\overline{F}_2$ is strictly plurisubharmonic at all points (except on A) on fibres sufficiently near to $\{s = 0\}$. Also since $a \neq 0$, for small values of s, $\{F_2\overline{F}_2 < \varepsilon\}$ will be relatively compact.

The second case is where $q = A_i \cap A_j$ is a singular point. Find $f_1 \in \Gamma(A_i, \underset{\nu}{\otimes} \mathcal{G}_i^*)$ such that f_1 vanishes to exactly first order at q and f_1 vanishes at all other $A_i \cap A_k$. Extend f_1 by 0 to be a section of $\underset{\nu}{\otimes} G^*$. Similarly find $f_2 \in \Gamma(A_j, \underset{\nu}{\otimes} \mathcal{G}_j^*)$ which vanishes to exactly first order at q and extend by 0. Find $f_3 \in \Gamma(A_i, \underset{\nu}{\otimes} \mathcal{G}_i^*)$ with $f_3(q) \neq 0$ and $f_4 \in \Gamma(A_j, \underset{\nu}{\otimes} \mathcal{G}_j^*)$ with $f_4(q) \neq 0$. We may require that f_3 and f_4 vanish at all other $A_i \cap A_k$ and $A_j \cap A_k$. A suitable linear multiple of f_4 will equal $f_3(q)$ at q. The section over $A_i \cup A_j$ given by f_3 and f_4

may be extended by 0 to give $F_3 \epsilon \Gamma(A, \underset{\nu}{\otimes} \mathcal{G}^*)$. If s is an appropriate A_i coordinate, u an appropriate A_j coordinate and t the fibre coordinate, then at q, F_1, F_2 and F_3 may be written locally as $F_1 = st^\nu$, $F_2 = ut^\nu$, $F_3 = t^\nu(a + a_1 s + a_2 s^2 + \ldots + b_1 u + b_2 u^2 + \ldots)$, $a \neq 0$. By adding appropriate multiples of F_1 and F_2 to F_3, we may assume that $a_1 = b_1 = 0$. An easy computation of the complex Hessian in (u,s,t) coordinates then shows that $F_1\bar{F}_1 + F_2\bar{F}_2 + F_3\bar{F}_3$ is strictly plurisubharmonic at all points (except on A) on fibres near $\{s = u = 0\}$.

Since A is compact, a finite number of F_i constructed in the above manner will suffice to make $\Sigma\, F_i\bar{F}_i$ strictly plurisubharmonic and $\{\Sigma\, F_i\bar{F}_i < \varepsilon\}$ relatively compact for all ε. A is then exceptional by Theorem 4.8. ∎

Applying Lemma 4.11 to the case at hand, we can find a neighborhood W of A in G and a holomorphic map $\chi: W \to C^n$ such that χ restricted to W – A is biholomorphic and $\chi(A) = 0$. Let $C^n = \{(z_1, \ldots, z_n)\}$ and $\chi(x) = (z_1, \ldots, z_n)$. The function $\chi^*(|z_1|^2 + \ldots + |z_n|^2) = P(x)$ is strictly plurisubharmonic on W – A. Recall that $A \subset M$ and V^* is a line bundle over M.

LEMMA 4.12. *Let* $\chi: W \to C^n$ *be as above. There exists a neighborhood* U *of* W – A *in* $V^* - M$ *and a strictly plurisubharmonic function* Q *in* U *such that* $Q|W - A = P$.

Remark: Lemma 4.12 is true in much greater generality than it is stated. Since we only need Lemma 4.12, we shall not consider the general case.

Proof: Let $\mathcal{J}$ be the sheaf of germs of holomorphic functions on $V^* - M$ which vanish on W – A. $\mathcal{J}$ is locally generated by a single function f_i. At each $x \in W - A$ choose a neighborhood U(x) of x in V^* such that f_i generates $\mathcal{J}$ at each point of U(x). Let

$$\phi_i = f_i : U(x) \to C. \quad \phi_i\,(W - A) = 0.$$

We may also choose U(x) suitably small so that the functions in $\chi: (W-A) \cap U(x) \to C^n$ extend to holomorphic functions on all of U(x) and thus give a map $\hat{\chi}_i: U(x) \to C^n$ which extends χ. $(\hat{\chi}_i, \phi_i): U(x) \to C^n \times C$ is then biholomorphic in some neighborhood N(x) of x. To see this, we have two cases, as illustrated in Figure 4.3 below.

(4.3)

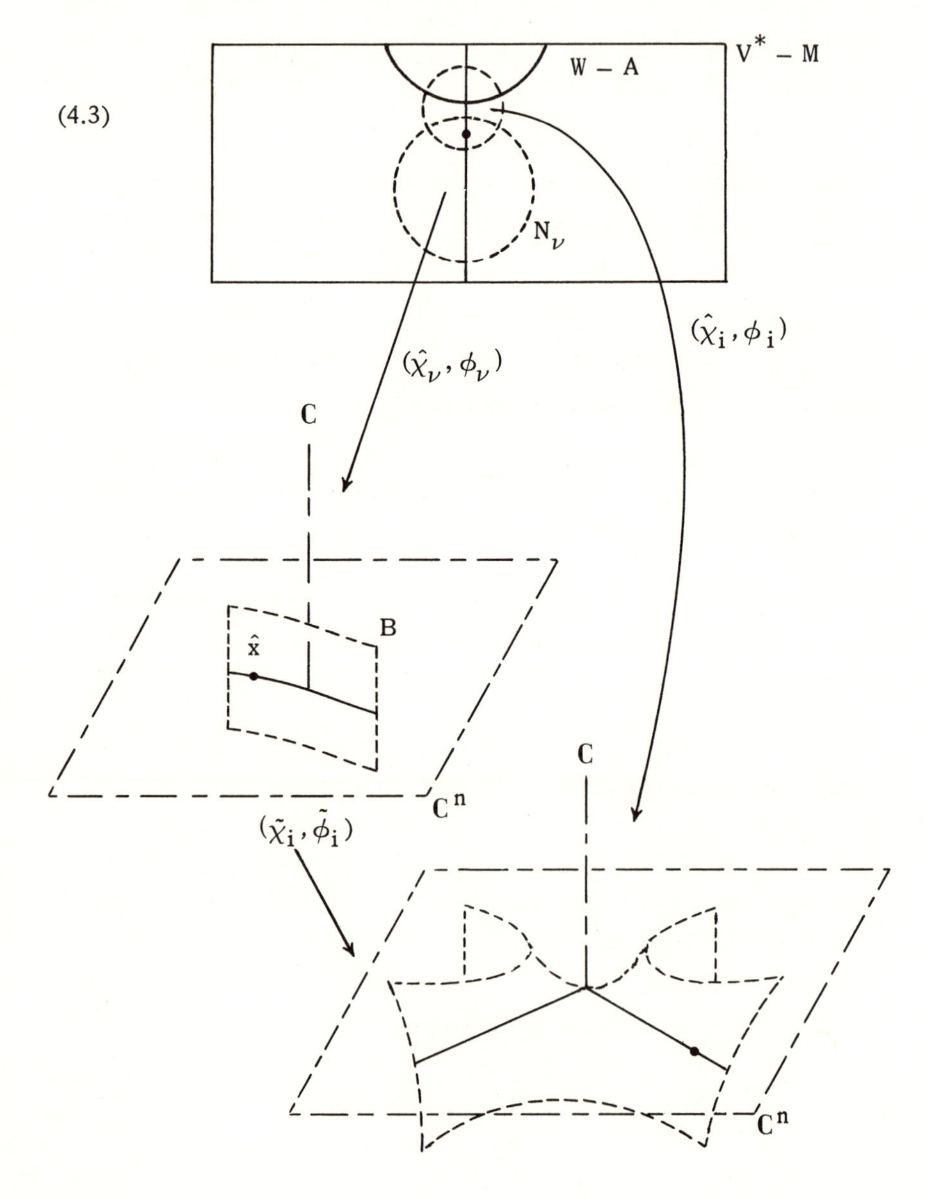

If x is a manifold point of $W - A$, f_i may be chosen as a coordinate function in $V^* - M$. Thus $(\hat{\chi}_\nu, \phi_\nu)$ will be biholomorphic near x and $(\hat{\chi}_\nu, \phi_\nu)$ $(W-A)$ will meet $C^n \times 0$ transversely. If x is a singular point, the tangential dimension $\dim t_x (W - A) = 3 = \dim M$. By Theorem V.B. 16 of G & R, $\hat{\chi}$ is already biholomorphic in some neighborhood of x.

Since $V^* - M$ is paracompact, there is a locally finite covering $\mathfrak{N} = \{N_i\}$ of some neighborhood of $W - A$ such that on N_i, there is a $(\hat{\chi}_i, \phi_i) : N_i \to C^n \times C = (z, y_i)$ which is biholomorphic. By construction, $\hat{\chi}_i | (W-A) \cap N_i = \chi | (W-A) \cap N_i$. Since f_i generates $\mathcal{J}$, $\phi_i(x) = 0$ if and only if $x \in W - A$. Let h_i be a C^∞ partition of unity subordinate to $\mathfrak{N}$. On $(W-A) \cap N_i$, $P(x) = |z|^2 \circ \hat{\chi}_i$. With k_i constants to be chosen later, we let

$$Q(x) = \sum_i h_i \left(|z|^2 \circ \hat{\chi}_i + k_i (|y_i|^2 \circ \phi_i) \right)$$

$$= \sum_i h_i \left(|z|^2 \circ \hat{\chi}_i + k_i |f_i|^2 \right)$$

In $W - A$, $f_i = 0$ so $Q(x) = P(x)$ for $x \in W - A$. We want to show that Q is strictly plurisubharmonic near $W - A$ for sufficiently large choice of k_i.

Look at $Q(x)$ near a point $x_o \in W - A$. $h_\nu(x_o) > 0$ for some ν. We want to work in $C^n \times C$ near $(\hat{\chi}_\nu, \phi_\nu)(x_o) = \hat{x}$. $(\tilde{\chi}_i, \tilde{\phi}_i) : (z, y_\nu) \to (z, y_i)$ should be a holomorphic map making (4.3) commutative on D, a small neighborhood of x_o. $(\tilde{\chi}_i, \tilde{\phi}_i) : C^n \times C$ should also be the identity on $C^n \times 0$. Letting $B = (\hat{\chi}_\nu, \phi_\nu)(D)$, commutativity requires that $(\tilde{\chi}_i, \tilde{\phi}_i)|_B = (\hat{\chi}_i, \phi_i) \circ (\hat{\chi}_\nu, \phi_\nu)^{-1}$. So we must extend the holomorphic functions defining $(\tilde{\chi}_i, \tilde{\phi}_i)|_B$ to holomorphic functions in some neighborhood of B such that $(\tilde{\chi}_i, \tilde{\phi}_i)|_{C^n \times 0}$ is the identity. We have already that $(\tilde{\chi}_i, \tilde{\phi}_i)|_{(C^n \times 0) \cap B}$ is the identity since $(C^n \times 0) \cap B$ is the image of

$(W-A) \cap D$. The arguments for $\tilde{\chi}_i$ and $\tilde{\phi}_i$ are similar. We shall thus only discuss one coordinate function $\tilde{z}_1(z_1, \ldots, z_n, y_\nu)$ of $\tilde{\chi}_i$. $\tilde{z}_1 | (C^n \times 0) \cap B = z_1$. Consider the function $\tilde{z}_1 - z_1$. It suffices to show that $\tilde{z}_1 - z_1$ has an extension that vanishes on $C^n \times 0$. In the case where x_o is a manifold point, since B meets $C \times 0$ transversely, we can just choose coordinates for $C^n \times C$ near $\hat{x}$ so that B and C^n are linear subspaces with coordinates (η_1, η_2, η_3) and $(\eta_2, \eta_3, \eta_4 \cdots \eta_{n+1})$. Now just write $\tilde{z}_1 - z_1$ in terms of (η_1, η_2, η_3). In the case where x_o is a singular point, let $(\eta_1, \eta_2, \ldots, \eta_n)$ be coordinates for C^n such that B may be given by $y_\nu = y_\nu(\eta_1, \eta_2, \eta_3)$. $\tilde{z}_1 - z_1$ vanishes on $(C^n \times 0) \cap B = B \cap \{y_\nu = 0\}$. $y_\nu = f_\nu(x)$ generates the ideal of $(\hat{\chi}_\nu, \phi_\nu)(W-A)$. Hence in (η_1, η_2, η_3) coordinates we may write

$$\tilde{z}_1 - z_1 = y_\nu(\eta_1, \eta_2, \eta_3)\, g\, (\eta_1, \eta_2, \eta_3)$$

for some holomorphic function g. In the ambient space $C^n \times C$, extend $\tilde{z}_1 - z_1$ by $y_\nu g(\eta_1, \eta_2, \eta_3)$, as a function of $(\eta_1, \eta_2, \ldots \eta_n, y_\nu)$. We may let $(\tilde{\chi}_\nu, \tilde{\phi}_\nu)$ be the identity everywhere.

We can now extend $(\hat{\chi}_\nu, \phi_\nu)^{-1} \circ Q(x)$, only defined on B, to a neighborhood of B in $C^n \times C$ by

$$Q(z, y_\nu) = \sum_i \tilde{h}_i \left(|z|^2 \circ \tilde{\chi}_i + k_i (|y_i|^2 \circ \tilde{\phi}_i)\right)$$

$$= \sum_i \tilde{h}_i \left(|z|^2 \circ \tilde{\chi}_i + k_i |\tilde{\phi}_i|^2\right)$$

where the $\tilde{h}_i$ are C^∞ functions such that $\tilde{h}_i|_B = h_i$ and $\Sigma\, \tilde{h}_i \equiv 1$. To show that Q(x) is strictly plurisubharmonic, it suffices to show that $\tilde{Q}(z, y_\nu)$ is strictly plurisubharmonic (for sufficiently large k_i). We will compute H, the complex Hessian of $\tilde{Q}$, on $B \cap (C^n \times 0)$ and show that H is positive definite there. Then H is positive definite in a neighborhood of $B \cap (C^n \times 0)$ and $\tilde{Q}$ is strictly plurisubharmonic, as desired. $(\tilde{\chi}_i, \tilde{\phi}_i)|_{C^n \times 0}$ is the identity, so $\tilde{Q}(z,0) = |z|^2$. Hence $\dfrac{\partial^2 \tilde{Q}}{\partial z_i \partial z_j} = I$

where I is the identity $n \times n$ matrix. In computing $\dfrac{\partial^2}{\partial y_\nu \partial \bar{y}_\nu} (\sum_i \tilde{h}_i k_i |\phi_i|^2)$

on $(C^n \times 0) \cap B$, both ϕ_i and $\bar{\phi}_i$ vanish. In using Leibnitz's rule both ϕ_i and $\bar{\phi}_i$ must be differentiated in order to yield non-zero summands. Hence derivatives of $\tilde{h}_i$ do not enter. $\dfrac{\partial^2 k_i \phi_i \bar{\phi}_i}{\partial y_\nu \partial \bar{y}_\nu} \geq 0$ for any choice of k_i and $\dfrac{\partial^2 \phi_\nu \bar{\phi}_\nu}{\partial y_\nu \partial \bar{y}_\nu} = 1$. Hence $\dfrac{\partial^2 \tilde{Q}}{\partial y_\nu \partial \bar{y}_\nu}$ is of the form $\alpha(\chi) + S(\phi) + k_\nu h_\nu$

where $\alpha(\chi)$, from $\sum \tilde{h}_i (|z|^2 \circ \tilde{\chi}_i)$ does not depend on the k_i and

$S(\phi) \geq 0$ regardless of the k_i. For the mixed derivatives, we again look at $\dfrac{\partial^2}{\partial z_j \partial \bar{y}_\nu} (\sum \tilde{h}_i k_i |\phi_i|^2)$. $\phi_i = \bar{\phi}_i = 0$ so we cannot differentiate $\tilde{h}_i$ and get a non-zero summand. But $\phi_i(z_j, 0) \equiv 0$, so $\dfrac{\partial \phi_i}{\partial z_j} = 0$. Thus all

the summands are zero. Hence the k_i do not enter into

$\beta(\chi) = \left(\dfrac{\partial^2 \tilde{Q}}{\partial z_j \partial \bar{y}_\nu}\right)$. Hence $H = \begin{pmatrix} I & \beta(\chi) \\ \beta(\chi) & \alpha(\chi) + S(\phi) + k_\nu h_\nu \end{pmatrix}$ where $\alpha(\chi)$ and $\beta(\chi)$ are independent of the k_i. We may perform row and column operations, independent of k_i, so that H is negative definite if

$$\begin{pmatrix} I & 0 \\ 0 & \alpha'(\chi) + S(\phi) + k_\nu h_\nu \end{pmatrix}$$

is negative definite. For large enough k_ν, k_ν chosen independently of the other k_i, this last matrix is negative definite. Since $\mathfrak{N}$ is a locally finite cover and we may assume that the $\bar{N}_i$ are compact, each k_ν has only a finite number of conditions on it. Hence we may choose the k_i large enough so that Q(x) is negative definite. ∎

We can now complete the proof of Theorem 4.9 as follows: M is the 0 - section of the line bundle V^*. Sections of V^* correspond to meromorphic functions on M which have poles up to a certain order on the A_i. Thus the constant function $s(x) \equiv 1$ is a non-trivial holomorphic section s of V^* over M. As a section, $s \neq 0$ on $M - A$. $s = 0$ on A since m* is generated by functions having non-trivial poles on A.

We have the following picture, see (4.4), of the total space of V^*. On U we have the strictly plurisubharmonic function Q of Lemma 4.12. $W = U \cap G$. On $W - A$, $Q(x) \to 0$ as $x \to A$. K is a neighborhood of A in V^* such that $K \cap W$ is relatively compact in W. With ∂ meaning boundary, for sufficiently small $\varepsilon > 0$, $Q(x) > \varepsilon$ for $x \in \partial K \cap W$.

(4.4)

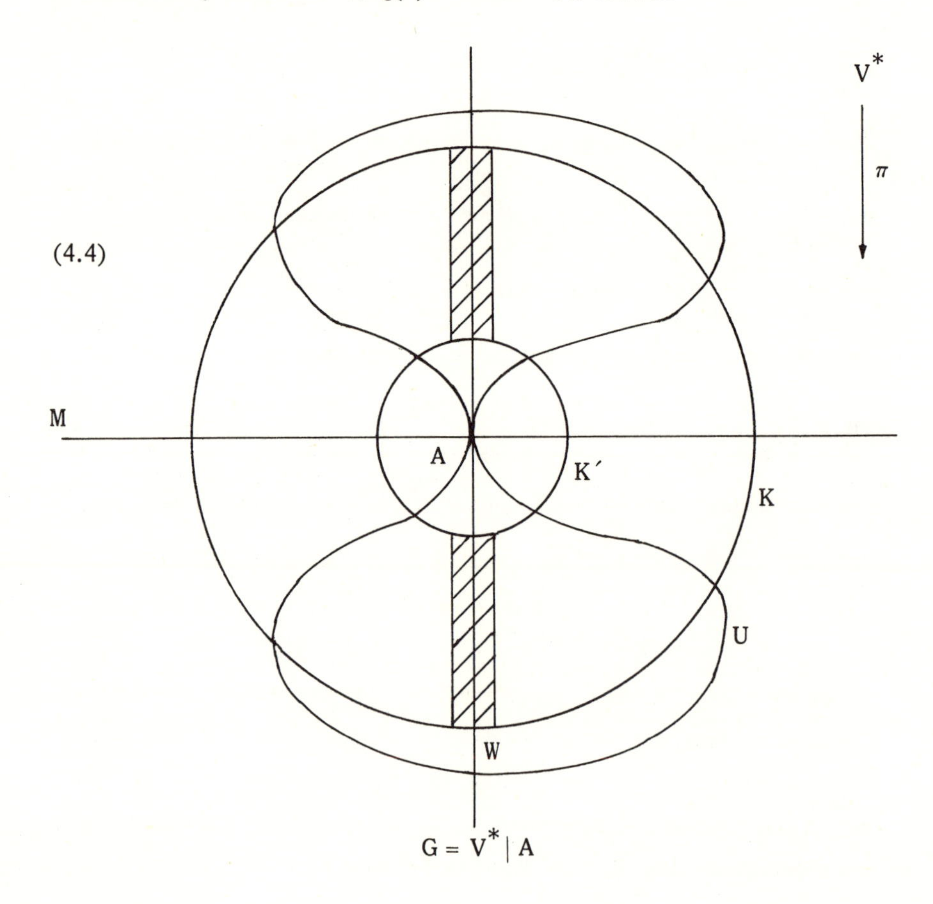

There exists a neighborhood K' of A in V^* such that K' is relatively compact in K and $Q(x) < \varepsilon$ for $x \in \partial K' \cap W$. Let $\pi: V^* \to M$ be the projection map in the line bundle V^*. Let $L \subset M$ be a sufficiently small neighborhood of A in M so that $(\overline{K} - K') \cap \pi^{-1}(L)$, which is shaded in (4.4), is relatively compact in U, $Q(x) > \varepsilon$ for $x \in \partial K \cap \pi^{-1}(L)$ and $Q(x) < \varepsilon$ for $x \in \partial K' \cap \pi^{-1}(L)$.

Choose k_o large enough so that for $k \geq k_o$, $k\, s(\partial L) \cap \overline{K} = \emptyset$. The relevant part of (4.4) now looks like (4.5).

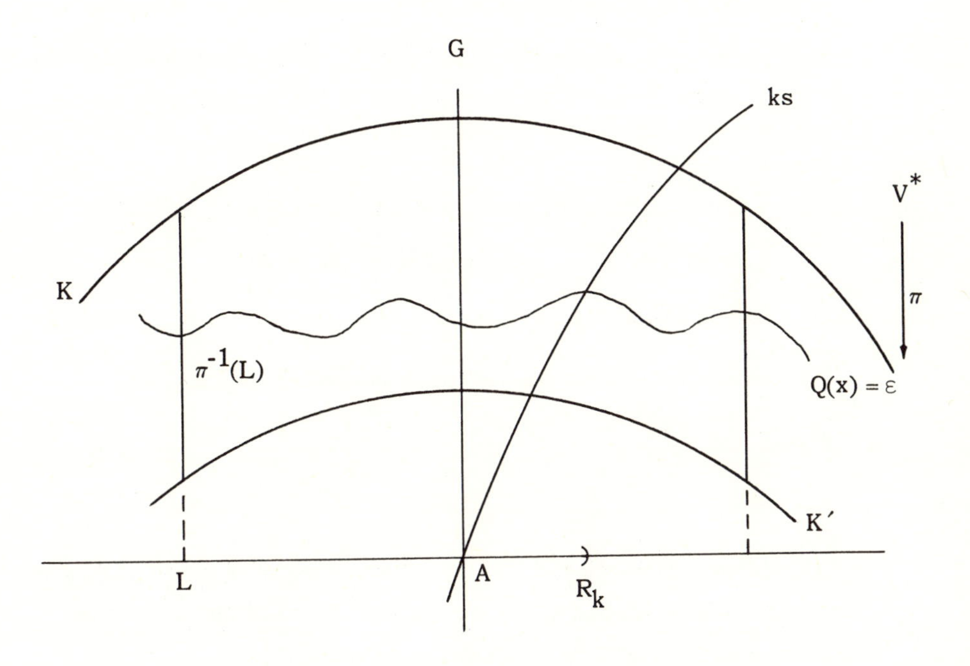

Let $R_k = \{x \in L \mid k\, s(x) \in \overline{K}' \text{ or } k\, s(x) \in K - \overline{K}' \text{ and } Q(k\, s(x)) < \varepsilon\}$. Since Q is strictly plurisubharmonic, so is $Q(k\, s(x))$ since s is a biholomorphic map onto its image. Hence R_k is a strictly Levi pseudoconvex neighborhood of A. R_k goes through arbitrarily small neighborhoods of A in M as $k \to \infty$. Hence by Theorem 4.8, A is an exceptional set in M.▌

CHAPTER V

MINIMAL RESOLUTIONS

Resolutions of normal singularities are not unique. We may always perform a quadratic transformation at a point of $\pi^{-1}(p)$. In this section we shall show that there is a unique minimal resolution. All other resolutions may be obtained from the minimal resolution by quadratic transformations.

We first need some machinery of a general nature about sheaves.

PROPOSITION 5.1. *(Mayer-Vietoris) Let* A *and* B *be open sets in the topological space* X *and* $\mathcal{S}$ *a sheaf over* X. *Then the following sequence is exact.*

$$0 \to \Gamma(A \cup B, \mathcal{S}) \xrightarrow{\iota} \Gamma(A,\mathcal{S}) \oplus \Gamma(B,\mathcal{S}) \xrightarrow{\rho} \Gamma(A \cap B,\mathcal{S}) \xrightarrow{\delta}$$
$$H^1(A \cup B, \mathcal{S}) \xrightarrow{\iota} \dots \to H^i(A \cup B,\mathcal{S}) \xrightarrow{\iota} H^i(A, \mathcal{S}) \oplus H^i(B,\mathcal{S})$$
$$\xrightarrow{\rho} H^i(A \cap B,\mathcal{S}) \xrightarrow{\delta} \dots$$

ι *is induced by* $\iota(\alpha) = \alpha \oplus \alpha$ *and* ρ *is induced by* $\rho(\beta \oplus \gamma) = \beta - \gamma$.

Proof: Let

$$0 \to \mathcal{S} \to \mathcal{C}^o \xrightarrow{\delta} \mathcal{C}^1 \xrightarrow{\delta} \mathcal{C}^2 \xrightarrow{\delta} \dots$$

be the canonical resolution of $\mathcal{S}$, i.e. $\mathcal{C}^o$ is the sheaf of germs of discontinuous sections of $\mathcal{S}$, and $\mathcal{C}^i$ is the sheaf of germs of discontinuous sections of $\mathcal{C}^{i-1}/\delta(\mathcal{C}^{i-2})$. For any open set D, $D \subset X$,

$$H^i(D,\mathcal{S}) = \ker \delta : \Gamma(D,\mathcal{C}^i) \to \Gamma(D,\mathcal{C}^{i+1})/\delta\Gamma(D,\mathcal{C}^{i-1}).$$

$$(5.1) \qquad 0 \to \Gamma(A\cup B,\mathcal{C}^i) \overset{\iota}{\to} \Gamma(A,\mathcal{C}^i) \oplus \Gamma(B,\mathcal{C}^i) \overset{\rho}{\to} \Gamma(A\cap B,\mathcal{C}^i) \to 0$$

where $\iota(\alpha) = \alpha \oplus \alpha$, after restriction, and $\rho(\beta \oplus \gamma) = \beta - \gamma$, after restriction to $A \cap B$, can easily be seen to be exact as follows: ι is an injection for if a section $\alpha \in \Gamma(A\cup B,\mathcal{C}^i)$ vanishes on both A and B, α vanishes on $A \cup B$ since sections are defined locally. Given $\eta \in \Gamma(A\cap B, \mathcal{C}^i)$, we may extend η by 0 to get $\tilde{\eta}$, a discontinuous section of $\mathcal{C}^{i-1}/\delta(\mathcal{C}^{i-2})$ on A. Then $\rho(\tilde{\eta} \oplus 0) = \eta$. Finally $\rho \circ \iota = 0$ and if $\rho(\beta \oplus \gamma) = \beta - \gamma = 0$ on $A \cap B$, then $\beta = \gamma$ on $A \cap B$. Then there is a section on $A \cup B$ which agrees with β and γ on A and B respectively. (5.1) thus gives a short exact sequence of complexes. The proposition then follows in the usual manner. ∎

DEFINITION 5.1. *Let $\Phi : G \to Y$ be a continuous map between the paracompact Hansdorff spaces G and Y. Let $\mathcal{S}$ be a sheaf over G. $\Phi_*(\mathcal{S})$, the direct image sheaf of $\mathcal{S}$ under Φ, is the sheaf over Y given by the following complete presheaf. If U is an open subset of Y,* $\Gamma(U,\Phi_*(\mathcal{S})) = \Gamma(\Phi^{-1}(U),\mathcal{S})$.

The verification that we do have a complete presheaf is straightforward, using the fact that the presheaf of $\mathcal{S}$ is complete.

We have, in fact, been implicitly using the notion of direct image sheaves. If $\pi : Y \to V$ is the normalization of an analytic space V, then $\pi_*(\mathcal{O}) = \tilde{\mathcal{O}}$, the sheaf of germs of weakly holomorphic functions on V. In Definition 4.2 one of the conditions for $\Phi : G \to Y$ to exhibit $A \subset G$ as exceptional is that $\Phi_*(\mathcal{O}) = \mathcal{O}$. As another example, if m is the ideal sheaf of A in G, $\Phi_*(m)$ is the ideal sheaf of $\Phi(A)$.

A very powerful, and difficult to prove, theorem of Grauert, [Gr2], [K], says that if $\Phi : G \to Y$ is a proper holomorphic map between the analytic spaces G and Y and if $\mathcal{S}$ is a coherent sheaf over G, then $\Phi_*(\mathcal{S})$ is a coherent sheaf over Y. Fortunately, ad hoc methods will suffice to verify this theorem in all the cases that we need. We shall thus avoid the use of Grauert's theorem. Observe that the definition of $\Phi: G \to Y$ exhibiting $A \subset G$ as exceptional insures that $\Phi_*(\mathcal{O}) = \mathcal{O}$ is coherent.

LEMMA 5.2. *Let* $\Phi: G \to Y$ *exhibit* $A \subset G$ *as an exceptional set. Let* $\mathcal{S}$ *be a coherent sheaf over* G *such that* $\Phi_*(\mathcal{S})$ *is coherent. Let* $A = \bigcup_{i=1}^{n} A_i$ *be the decomposition of* A *into irreducible components. Let* p_i *be the ideal sheaf of* A_i. *For any non-negative integers* r_i, $\Phi_*(p_1^{r_1} \cdots p_n^{r_n} \mathcal{S})$ *is coherent.*

Let $f_1, \ldots, f_t$ *be holomorphic functions on* Y *with* $\Phi(A) = V(f_1, \ldots, f_t)$. *Let* $\tilde{f}_i = \Phi_*(f_i)$ *and let* $\mathcal{I}$ *be the ideal sheaf generated by the* $\tilde{f}_i$. *Then* $\Phi_*(\mathcal{I}\mathcal{S})$ *is coherent.*

Proof: Let $\mathcal{J}$ denote either $p_1^{r_1} \cdots p_n^{r_n}$ or $\mathcal{I}$. The beginning of the proof is identical in both cases.

$\Phi_*(\mathcal{J}\mathcal{S}) \approx \Phi_*(\mathcal{S})$ off $\Phi(A)$ so that $\Phi_*(\mathcal{J}\mathcal{S})$ is coherent off $\Phi(A)$. Also $\Phi_*(\mathcal{J}\mathcal{S}) \subset \Phi_*(\mathcal{S})$ so that, since the image of an $\mathcal{O}$-module map between coherent sheaves is coherent, it suffices to show that $\Phi_*(\mathcal{J}\mathcal{S})$ is finitely generated near a point $p \in \Phi(A)$. $\Phi_*(\mathcal{J}\mathcal{S})_p \subset \Phi_*(\mathcal{S})_p$, a finitely generated module over the Noetherian ring $\mathcal{O}_p$. Hence there exist germs of sections $g_1, \ldots, g_s \in \Phi_*(\mathcal{J}\mathcal{S})_p$ which generate $\Phi_*(\mathcal{J}\mathcal{S})_p$. $g_1, \ldots, g_s$ need not, however, generate $\Phi_*(\mathcal{J}\mathcal{S})_q$ for q near p so that we must add additional generators. For $q \neq p$ and q near p, $\Phi_*(\mathcal{J}\mathcal{S})_q \approx \Phi_*(\mathcal{S})_q$. $\Phi_*(\mathcal{S})$ is coherent, so there exist germs of sections $h_1, \ldots, h_w \in \Phi_*(\mathcal{S})_p$ which generate $\Phi_*(\mathcal{S})_q$ for q near p. We need, of course, for $h_1, \ldots, h_w$ to be in $\Phi_*(\mathcal{J}\mathcal{S})$. Consider first the case that $\mathcal{J} = \mathcal{I}$. $f_j h_k \in \Phi_*(\mathcal{I}\mathcal{S})$ for all f_j and h_k. Since p is the locus of the common zeroes of the f_i, $\{f_j h_k\}$ generate $\Phi_*(\mathcal{S})_q$ for $q \neq p$ and hence $\Phi_*(\mathcal{I}\mathcal{S})_q$ for $q \neq p$. Thus $\{g_1, \ldots g_s, \ldots, f_j h_k, \ldots\}$ is a finite set of generators for $\Phi_*(\mathcal{I}\mathcal{S})$ in some neighborhood of p. Hence $\Phi_*(\mathcal{I}\mathcal{S})$ is coherent.

Now consider $\mathcal{J} = p_1^{r_1} \cdots p_n^{r_n}$. Let $f_1, \ldots, f_t$ be holomorphic functions near p with $p = V(f_1, \ldots, f_t)$. $\Phi^*(f_j)$ vanishes on A_i, for all i. Hence $\Phi^*(f_j^r)$ is a section of $p_1^{r_1} \cdots p_n^{r_n}$ for $r = \max(r_i)$. $\{f_j^r\}$ still have p as their only common zero. Hence $\{f_j^r h_k\}$ generate $\Phi_*(\mathcal{J}\mathcal{S})_q$ for $q \neq p$. Hence, as before, $\Phi_*(\mathcal{J}\mathcal{S})$ is coherent.∎

LEMMA 5.3. *Let* $\Phi: G \to Y$ *represent* $A \subset G$ *as an exceptional set. Let* $U \subset W$ *be neighborhoods of* A *such that* $\Phi(U)$ *and* $\Phi(W)$ *are Stein spaces. Let* $\mathcal{S}$ *be a coherent sheaf over* G *such that* $\Phi_*(\mathcal{S})$ *is coherent. Then the restriction map* $r: H^i(W, \mathcal{S}) \to H^i(U, \mathcal{S})$ *is an isomorphism for all* $i \geq 1$.

Proof: $\Phi(U)$ and $\Phi(W)$ are open. Using the Mayer-Vietoris sequence, we have the following commutative diagram with exact rows.
$W = (W-A) \cup U$ and $U - A = (W-A) \cap U$.

$$\begin{array}{ccccc} 0 \to \Gamma(\Phi(W), \Phi_*(\mathcal{S})) & \xrightarrow{\iota} & \Gamma(\Phi(U), \Phi_*(\mathcal{S})) \oplus \Gamma(\Phi(W-A), \Phi_*(\mathcal{S})) \\ \downarrow \Phi^* & & \downarrow \Phi^* \\ 0 \to \Gamma(W, \mathcal{S}) & \xrightarrow{\iota'} & \Gamma(U,\mathcal{S}) \oplus \Gamma(W-A, \mathcal{S}) \end{array} \tag{5.2}$$

$$\begin{array}{ccc} \xrightarrow{\rho} \Gamma(\Phi(U-A), \Phi_*(\mathcal{S})) \to & & H^1(\Phi(W), \Phi_*(\mathcal{S})) \to \ldots \\ \downarrow \Phi^* & & \downarrow \Phi^* \\ \xrightarrow{\rho'} \Gamma(U-A, \mathcal{S}) \to & & H^1(W, \mathcal{S}) \to \ldots \end{array}$$

Φ^* is induced by the definition $\Gamma(U, \Phi_*(\mathcal{S})) = \Gamma(\Phi^{-1}(U), \mathcal{S})$, which also induces a map on cochains and hence on cohomology. $H^1(\Phi(W), \Phi_*(\mathcal{S})) = 0$ by Cartan's Theorem B. Hence ρ is surjective and ρ' is surjective. Thus again using Theorem B, we may replace (5.2) by

$$\begin{array}{ccccc} 0 \to & 0 & \xrightarrow{\iota} & 0 \oplus H^1(\Phi(W-A), \Phi_*(\mathcal{S})) & \xrightarrow{\rho} \\ & \downarrow \Phi^* & & \downarrow \Phi^* & \\ 0 \to & H^1(W, \mathcal{S}) & \xrightarrow{\iota'} & H^1(U,\mathcal{S}) \oplus H^1(W-A, \mathcal{S}) & \xrightarrow{\rho'} \end{array}$$

$$\begin{array}{ccccc} H^1(\Phi(U-A), \Phi_*(\mathcal{S})) & \to & 0 & \to & \dots \\ \downarrow \Phi^* & & \downarrow & & \\ H^1(U-A, \mathcal{S}) & \to & H^2(W', \mathcal{S}) & \to & \dots \end{array} \tag{5.3}$$

But $\Phi^* : H^i(\Phi(W-A), \Phi_*(\mathcal{S})) \to H^i(W-A, \mathcal{S})$ and $\Phi^*: H^i(\Phi(U-A), \Phi_*(\mathcal{S})) \to H^i(U-A, \mathcal{S})$ are isomorphisms since Φ is an isomorphism off A. An easy diagram chase of (5.3) now shows that $r: H^i(W,\mathcal{S}) \to H^i(U, \mathcal{S})$ is an isomorphism for $i \geq 1$. ∎

The following theorem will prove to be extremely useful.

THEOREM 5.4. *Let* $\Phi: W \to Y$ *represent* $A \subset W$ *as an exceptional set with* W *a manifold and* Y *a Stein space. Let* A *be everywhere of codimension* 1, *and let* m *be the ideal sheaf of* A. *Let* $\mathcal{F}$ *be a coherent subsheaf of a free sheaf on* W *such that* $\Phi_*(\mathcal{F})$ *is coherent. Then there exists a positive integer* k *such that the map* $H^\nu(W,\mathcal{F} \cdot m^k) \to H^\nu(W,\mathcal{F})$, $\nu = 1,2,3,\dots$, *induced by multiplication* $\mathcal{F} \cdot m^k \to \mathcal{F}$ *is the zero map.*

Proof: Let $A = \bigcup_{i=1}^{n} A_i$ be the decomposition of A into irreducible components. If p_i is the ideal sheaf of A_i, then locally p_i is generated by a single function [G & R, Lemma VIII.B.12]. Thus $m = p_1 \cdots p_n$ is the ideal sheaf of A by Theorem II. E.19 of G & R. Also, $m^k = p_1{}^k \cdots p_n{}^k$. Then by Lemmas 5.2 and 5.3, if U is a relatively compact subset of W such that $\Phi(U)$ is Stein, the restriction map from W to U induces an isomorphism on the higher cohomology groups of $\mathcal{F} \cdot m^k$.

Let $f_1, \dots, f_t$ be holomorphic functions near $\Phi(A)$ with $\Phi(A)$ as their common zero. We may choose U so that $f_1, \dots, f_t$ are all defined on $\Phi(U)$. Let n be the ideal sheaf on U generated by $\Phi^*(f_1), \dots, \Phi^*(f_t)$. $\Phi_*(n)$ is coherent by Lemma 5.2. $m \supset n$ and if r is the maximum to which some f_i vanishes on some A_j, $m^r \subset n$. Let $\mathcal{F}^t$ denote $\mathcal{F} \oplus \dots \oplus \mathcal{F}$, with t summands. [For an ideal sheaf $\mathcal{I}$, $\mathcal{I}^t$ will continue to denote $\mathcal{I} \cdots \mathcal{I}$. The meaning should be clear from the context.] Let $\phi: \mathcal{F}^t \to \mathcal{F} \cdot n$ be given by

$\phi(g_1,\ldots,g_t) = \Sigma\, g_i f_i$. ϕ is surjective by the definition of $\mathcal{F}\cdot n$. Let $\mathcal{R} = \ker \phi$.

$$0 \to \mathcal{R} \xrightarrow{\iota} \mathcal{F}^t \xrightarrow{\phi} \mathcal{F}\cdot n \to 0 \tag{5.4}$$

is then an exact sheaf sequence. $\mathcal{R}$ satisfies the hypotheses of this theorem. It is a coherent subsheaf of $\mathcal{F}^t$, which in turn is a subsheaf of a free sheaf. For any open set Q,

$$0 \to \Gamma(Q,\mathcal{R}) \to \Gamma(Q,\mathcal{F}^t) \to \Gamma(Q,\mathcal{F}\cdot n)$$

is exact. $\Phi_*(\mathcal{R}) = \ker \phi_* : \Phi_*(\mathcal{F}^t) \to \Phi_*(\mathcal{F}\cdot n)$ is coherent since it is the kernel of an $\mathcal{O}$-module sheaf map of coherent sheaves.

Multiplying (5.4) by m^s gives

$$0 \to \mathcal{R}\cdot m^s \xrightarrow{\iota_*} (\mathcal{F}\cdot m^s)^t \xrightarrow{\phi_*} \mathcal{F}\cdot m^s n \to 0 \tag{5.5}$$

By Lemma 5.2, the sheaves in (5.5) satisfy the hypotheses of the theorem. To see that (5.5) is exact at $x \in W$, we argue as follows. m^s is locally generated by a single function h, so in deriving (5.5) from (5.4) we only have to multiply by this function h. Suppose $\iota_*(hs)_x = h\,\iota(s)_x = 0$. Since W is a manifold, $\mathcal{O}_x$ is an integral domain. $h_x \neq 0$ so $\iota(s)_x = 0$. From the exactness of (5.4), $s_x = 0$ so $(hs)_x = 0$. Similar arguments complete the verification that (5.5) is exact.

If k is a smaller integer than s, (5.5) yields the following commutative diagram. The rows are exact.

$$\begin{array}{ccccc} H^\nu(U,(\mathcal{F}\cdot m^s)^t) & \to & H^\nu(U,\mathcal{F}\cdot m^s n) & \to & H^{\nu+1}(U,\mathcal{R}\cdot m^s) \\ \downarrow \alpha & & \downarrow \beta & & \downarrow \gamma \\ H^\nu(U,(\mathcal{F}\cdot m^k)^t) & \xrightarrow{\phi_*} & H^\nu(U,\mathcal{F}\cdot m^k n) & \to & H^{\nu+1}(U,\mathcal{R}\cdot m^k) \end{array} \tag{5.6}$$

Since U is relatively compact and locally Stein near the boundary, it has a finite Leray cover. Hence $H^\nu(U,\mathcal{S}) = 0$ for all sufficiently large ν

and any coherent sheaf $\mathcal{S}$. Thus we can start to prove Theorem 5.4 by decreasing induction on ν, knowing that it holds for all sufficiently large ν.

Let Im* denote the image in $H^\nu(U,\mathcal{F})$. Then as k increases, $\mathrm{Im}^*H^\nu(U,\mathcal{F}\cdot m^k)$ decreases. But $H^\nu(U,\mathcal{F})$ is a finite dimensional vector space by Lemma 5.2 and Theorem IX. B.9 of G & R. Hence for sufficiently large k_o, if $k \geq k_o$, $\mathrm{Im}^*H^\nu(U,\mathcal{F}\cdot m^k) = \mathrm{Im}^*H^\nu(U,\mathcal{F}\cdot m^{k+1})$. Since $m \supset n \supset m^r$,

$$\mathrm{Im}^*H^\nu(U,\mathcal{F}\cdot n\cdot m^s) = \mathrm{Im}^*H^\nu(U,\mathcal{F}\cdot n\cdot m^k) = \mathrm{Im}^*H^\nu(U,\mathcal{F}\cdot m^k)$$

for $k \geq k_o$. By the induction hypothesis on $\nu+1$, we may choose s sufficiently larger than k_o so that in (5.6) γ is the zero map when $k = k_o$. Exactness in the rows of (5.6) then implies that $\mathrm{im}\,\beta \subset \mathrm{im}\,\phi_*$. Let $\xi_1,\ldots,\xi_\tau$ be cocycles in $\mathcal{F}\cdot m^s\cdot n$ such that $\mathrm{cls}[\xi_1],\ldots,\mathrm{cls}[\xi_\tau]$ are a basis of $\mathrm{Im}^*H^\nu(U,\mathcal{F}\cdot m^s\cdot n) = \mathrm{Im}^*H^\nu(U,\mathcal{F}\cdot m^{k_o})$. Let $\mathrm{cls}[\zeta_1],\ldots,\mathrm{cls}[\zeta_\sigma]$ be a basis of the kernel of the map $H^\nu(U,\mathcal{F}\cdot m^{k_o}) \to H^\nu(U,\mathcal{F})$ so that $\mathrm{cls}[\xi_1],\ldots,\mathrm{cls}[\xi_\tau]$, $\mathrm{cls}[\zeta_1],\ldots,\mathrm{cls}[\zeta_\sigma]$ is a basis for $H^\nu(U,\mathcal{F}\cdot m^{k_o})$. Since $\mathrm{im}\,\beta \subset \mathrm{im}\,\phi_*$, there exist complex numbers u_{ijk} and $v_{i\ell k}$ such that in $H^\nu(U,\mathcal{F}\cdot m^{k_o}\cdot n)$

$$\mathrm{cls}[\xi_k] = \sum_i f_i \Big(\sum_j u_{ijk}\mathrm{cls}[\xi_j] + \sum_\ell v_{i\ell k}\mathrm{cls}[\zeta_\ell]\Big)$$

with $1 \leq i \leq t$, $1 \leq j,k \leq \tau$, $1 \leq \ell \leq \sigma$. In $H^\nu(U,\mathcal{F})$, $\mathrm{cls}[\zeta_\ell]$ is trivial. Thus in $H^\nu(U,\mathcal{F})$,

$$\mathrm{cls}[\xi_k] = \sum_{i,j} f_i u_{ijk}\mathrm{cls}[\xi_j] = \sum_{i,j} u_{ijk}\mathrm{cls}[f_i\xi_j]. \tag{5.7}$$

Now choose a small enough neighborhood N of A with $\Phi(N)$ Stein so that $\sum_{i,j} |u_{ijk}f_i(x)| < \frac{1}{2}$ for all $x \in N$. By Lemma 5.2, restriction to N is an isomorphism on cohomology, so (5.7) still holds. We shall now show that (5.7) implies that $\mathrm{Im}^* H^{\nu}(N, \mathcal{F} \cdot m^{k_o})$ is the zero subspace of $H^{\nu}(N, \mathcal{F})$. $H^{\nu}(N, \mathcal{F})$ has the natural structure of a linear topological space as follows. $\mathcal{F}$ is a Frechet sheaf by VIII.A.4 of G & R, so in particular $\Gamma(Q, \mathcal{F})$ is a Frechet space for any open subset Q of N. Let $\{N_1\}$ be a finite Leray covering of N. The cochains of the covering $\{N_1\}$ with values in $\mathcal{F}$ are then elements of a finite direct sum of Frechet spaces and therefore also form a Frechet space. The coboundary operator δ maps C^{μ} to $C^{\mu+1}$ where C^{μ} are the μ-cochains. δ is continuous.

$$H^{\nu}(N, \mathcal{F}) = \frac{\ker\ \delta : C^{\nu} \to C^{\nu+1}}{\mathrm{im}\ \ \delta : C^{\nu-1} \to C^{\nu}}$$

Since δ is continuous, ker δ is a closed subspace of C^{ν} and hence also a Frechet space. $H^{\nu}(N, \mathcal{F})$ is of finite dimension λ. Let $\eta_1, \ldots, \eta_\lambda$ be cocycles in C^{ν} whose image in $H^{\nu}(N, \mathcal{F})$ is a basis. Consider the linear map $\tilde{\delta} : C^{\nu-1} \oplus C^{\lambda} \to \ker \delta$ induced by $\tilde{\delta}(a \oplus \eta_i) = \delta a + \eta_i$. $\tilde{\delta}$ is continuous and onto and by the open mapping theorem is an open map. $\ker \tilde{\delta} = (\ker \delta : C^{\nu-1} \to C^{\nu}) \oplus 0$. Hence $\ker \delta \approx C^{\nu-1}/\ker \delta \oplus C^{\lambda}$ is a topological direct sum decomposition. In particular, $H^{\nu}(N, \mathcal{F})$ receives the usual Hausdorff topology on C^{λ}.

Now return to (5.7). The semi-norms on C^{λ} which define the topology on $H^{\nu}(N, \mathcal{F})$ are supremums over compact sets since $\mathcal{F}$ is a subsheaf of a free sheaf by assumption. Let $\|\ \|$ be any one such seminorm. Let $B = \max \|\xi_k\|$. Let $\xi'_k = \sum_{i,j} f_i u_{ijk} \xi_k$. Then since $\sum_{i,j} |u_{ijk} f(x)| < \frac{1}{2}$ for all x, $\|\xi'_k\| \leq \frac{1}{2} B$. Iterating this procedure, $\xi''_k = \sum_{i,j} f_i u_{ijk} \xi'_k$ has $\|\xi''_k\| \leq \frac{1}{2^2} B$. Thus $\xi_k, \xi'_k, \xi''_k, \ldots,$ all cocyles representing the same cohomology class, approach 0 in the cocycle space. Since the projection of the cocycles onto $H^{\nu}(N, \mathcal{F})$ is a continuous map, $\operatorname{cls}[\xi_k]$ is arbitrarily close to 0. But, as shown above, $H^\nu(N,\mathcal{F})$ has a Hausdorff topology. Hence $\operatorname{cls}[\xi_k] = 0$ for all k and the induction step is proved. ∎

THEOREM 5.5. *Let* A *be a Riemann surface of genus* 0 *which is non-singularly embedded in the* 2*-dimensional manifold* M. *If* $A \cdot A = -1$, *then* A *is the result of a quadratic transformation, i.e. there is a* 2*-dimensional manifold* M' *and a point* $p \in M'$ *such that if* $\pi: M'' \to M'$ *is the quadratic transformation at the point* p, *then there is a biholomorphic map* $\phi: M \to M''$ *such that* $\phi(A) = \pi^{-1}(p)$.

Proof: By Theorem 4.9, A is exceptional in M. Hence there is a relatively compact strictly pseudoconvex neighborhood U of A such that A is the maximal compact subset of U and $\Phi: U \to Y$, exhibiting A as exceptional, has Y a Stein space.

LEMMA 5.6. *Let* A *be as in Theorem 5.5. Let* $\mathcal{I}$ *be the ideal sheaf of* A. *Let* U *be as above. Then for* $t \geq 0$, $H^1(U,\mathcal{I}^t) = 0$.

Proof: For $r \geq t$, we have the exact sheaf sequence

$$0 \to \mathcal{I}^{r+1} \to \mathcal{I}^r \to \mathcal{I}^r/\mathcal{I}^{r+1} \to 0. \tag{5.8}$$

(5.8)yields exact cohomology sequence,

$$H^1(U,\mathcal{I}^{r+1}) \xrightarrow{\alpha_r} H^1(U,\mathcal{I}^r) \to H^1(U,\mathcal{I}^r/\mathcal{I}^{r+1}). \tag{5.9}$$

As discussed just before Theorem 2.6, $\mathcal{I}^r/\mathcal{I}^{r+1})$ is supported on A and is the sheaf of germs of sections of $\underset{r}{\otimes} N^*$, where N is the normal bundle of the embedding of A. Thus

$$H^1(U,\mathcal{I}^r/\mathcal{I}^{r+1}) \approx H^1(A,\underset{r}{\otimes}\mathcal{N}^*).$$

By Serre duality $H'(A,\underset{r}{\otimes}\mathcal{N}^*)$ has the same dimension as does $\Gamma(A,\kappa\otimes(\underset{r}{\otimes}\mathcal{N}))$, where κ denotes both the canonical bundle and its sheaf of germs of sections. $c(\kappa) = -2$ since A is the projective line. Hence

$$c(\kappa \otimes (\underset{r}{\otimes} N)) = -2 - r < 0 \ ,$$

so that $H^1(U, \mathcal{J}^r / \mathcal{J}^{r+1}) = 0$. Hence α_r in (5.9) is onto. Composing the α_r, we see that $\phi: H^1(U, \mathcal{J}^s) \to H^1(U, \mathcal{J}^t)$ is onto for all $s \geq t$. By Theorem 5.4, ϕ is the zero map for sufficiently large s. Hence $H^1(U, \mathcal{J}^t) = 0$ as claimed.∎

Returning to the proof of Theorem 5.5, from (5.8) and Lemma 5.6, we see that the map $\Gamma(U, \mathcal{J}) \to \Gamma(U, \mathcal{J}/\mathcal{J}^2) \approx \Gamma(A, \mathcal{N}^*)$ is onto. By Riemann-Roch, there exist $f_1, f_2 \in \Gamma(A, \mathcal{N}^*)$ which are a basis of $\Gamma(A, \mathcal{N}^*)$. There exist F_1 and F_2 in $\Gamma(U, \mathcal{J})$ which restrict to f_1 and f_2 respectively. In a local coordinate system (x,y) with $A = \{y = 0\}$, we may expand F_1 and F_2 as

$$\begin{aligned} F_1(x,y) &= y\, f_{11}(x) + y^2 f_{12}(x) + \ldots = y[f_{11}(x) + y\, f_{12}(x) + \ldots \\ F_2(x,y) &= y\, f_{21}(x) + y^2 f_{22}(x) + \ldots = y[f_{21}(x) + y\, f_{22}(x) + \ldots \ . \end{aligned} \tag{5.10}$$

$f_{11}(x)$ and $f_{21}(x)$ represent f_1 and f_2 locally. f_1 and f_2 can have no common zero since otherwise a linear combination would have a zero of second order and hence, as a section of a bundle of Chern class 1, would be identically 0. Thus $f_{11}(x)$ and $f_{21}(x)$ have no common zero. Then in (5.10),

$$|f_{11}(x) + y\, f_{12}(x) + \ldots|^2 + |f_{21}(x) + y\, f_{22}(x) + \ldots|^2$$

is non-zero on A. Let U_ε be that connected component of

$$\{p \in M \mid \ |F_1(p)|^2 + |F_2(p)|^2 < \varepsilon\}$$

which contains A. Then for all sufficiently small $\varepsilon > 0$, U_ε is relatively compact in M.

$$F = (F_1, F_2) : U_\varepsilon \to B_\varepsilon = \{(z_1,z_2) \mid |z_1|^2 + |z_2|^2 < \varepsilon\}$$

is then a proper holomorphic map with $A = F^{-1}(0,0)$ for all sufficiently small $\varepsilon > 0$. Let us now calculate the Jacobian $J(x,y)$ of F near $(0,0)$. F and $(F_1, F_2 + \alpha F_1)$, $\alpha \in C$, have equal Jacobians. Thus we may assume that $f_{21}(0) = 0$. Then $f_{11}(0) \neq 0$ and $f_{21}'(0) \neq 0$ since f_1 and f_2 are a basis of $\Gamma(A, \mathfrak{N}^*)$.

$$J(x,y) = \begin{pmatrix} \dfrac{\partial F_1}{\partial x} & \dfrac{\partial F_2}{\partial y} \\ \\ \dfrac{\partial F_2}{\partial x} & \dfrac{\partial F_2}{\partial y} \end{pmatrix} = \begin{pmatrix} y\, f_{11}'(x) + \dots & f_{11}(x) + \dots \\ \\ y\, f_{21}'(x) + \dots & f_{21}(x) + \dots \end{pmatrix}$$

$$= y(f_{11}'(x) f_{21}(x) - f_{11}(x) f_{21}'(x)) + y^2(\quad) + \dots \quad .$$

At

$$x = 0,\ f_{11}'(0)\, f_{21}(0) - f_{11}(0)\, f_{21}'(0) = -f_{11}(0) f_{21}'(0) \neq 0.$$

Hence $J(x,y) \neq 0$ for all sufficiently small $y \neq 0$. We may then choose ε small enough so that on U_ε, $F\colon U_\varepsilon - A \to B_\varepsilon - (0,0)$ is a proper map which is a local homeomorphism. $F \mid U_\varepsilon - A$ is, from purely topological considerations, a covering map. (We must verify that each $q \in B_\varepsilon - (0,0)$ has a neighborhood T such that F is a homeomorphism on each component of $F^{-1}(T)$. Let $r_1, \dots, r_s = F^{-1}(q)$ and let $T_1, \dots, T_s$ be disjoint neighborhoods of $r_1, \dots, r_s$ respectively such that F is a homeomorphism on each T_i. There exists a neighborhood T of q such that $F^{-1}(T) \subset \cup T_1$ by Lemma 3.2.) $B_\varepsilon - (0,0)$ is simply connected since it has the homotopy type of S^3. Hence F is one-to one on $U_\varepsilon - A$ and hence biholomorphic on $U_\varepsilon - A$.

We establish the required isomorphism ϕ as follows. Let B' be the result of a quadratic transformation at $(0,0)$ in B_ε.

$(z_1,z_2) = (\xi\zeta, \zeta) = (\zeta', \xi'\zeta')$. Let $\sigma = \pi^{-1}(0,0)$, $\pi: B' \to B_\varepsilon$. The points of σ correspond to complex lines through (0,0) and may be given by the points (z_1,z_2) on these lines, as homogeneous coordinates for P^1. (f_1,f_2) is a well defined point in P^1 and it is easy to verify that the map $(f_1,f_2): A \to P^1$ is biholomorphic. Extend the map $F: U_\varepsilon - A \to B_\varepsilon - (0,0)$ to $\phi: U_\varepsilon \to B'$ by mapping q to $(f_1(q), f_2(q))$. Consider a point $q \in A$ where say $f_2(q) \neq 0$. ϕ may be given locally near q as follows. $(\xi, \zeta) = (z_1/z_2, z_2) =$

$$(F_1/F_2, F_2) = \left(\frac{f_{11}(x) + yf_{12}(x) + \dots}{f_{21}(x) + yf_{22}(x) + \dots},\ yf_{21}(x) + y^2 f_{22}(x) + \dots\right).$$

$f_{21}(p) \neq 0$ since $f_2(p) \neq 0$. Hence ϕ is holomorphic near q. Since ϕ is proper, ϕ^{-1} is continuous and hence holomorphic by the Riemann removable singularity theorem. To extend ϕ to a map $M \to M''$, we just patch $M-A$ to $B'-\sigma$ via $\phi: U_\varepsilon \to B'$.∎

Thus as mentioned in Chapter II, a P^1 with weight -1 in a weighted graph may be collapsed with the resulting space again a manifold.

The next theorems tell us in general which weighted graphs correspond to sets A which can be collapsed and still give manifolds.

THEOREM 5.7. *Let $\pi: M' \to M$ be a surjective proper holomorphic map between connected 2-dimensional manifolds. Suppose that there is a compact, proper subvariety $S \subset M$ such that π restricted to $M' - \pi^{-1}(S)$ is biholomorphic. Then there are a finite number of points $p_1, \dots, p_t$ in M such that π is obtained by a finite number of iterated quadratic transformations at the points $p_1, \dots, p_t$.*

Proof: $S' = \pi^{-1}(S)$ is a proper subvariety of M', which is connected. Hence S' is of dimension 1 or 0. Since S' is compact, it has only a finite number of irreducible 1-dimensional components. Thus there are only a finite number of points $p \in S$ such that $\pi^{-1}(p)$ is 1-dimensional. Suppose for a $p \in S$, $\pi^{-1}(p) = \{q_1, \ldots, q_s\}$ is discrete. There are disjoint neighborhoods N_i of the q_i with $\cup\, \pi(N_i)$ a neighborhood of p. Let N be a connected neighborhood of p with $N \subset \cup\, \pi(N_i)$. $N - S$ is connected by G & R, 1.C.4. Hence, since π is biholomorphic on $\pi^{-1}(N-S)$, $\pi^{-1}(N-S)$ is connected. Then there can be only one q_j. Thus except on the inverse image of a finite set, π is one-to-one. Since π is proper it is also a homeomorphism there. Since M' and M' are manifolds, π^{-1} is holomorphic on the complement of this finite set by the Riemann removable singularity theorem. Therefore in the hypotheses of our theorem, we may assume that S is a finite point set, $p_1, \ldots, p_t$. It is now a local theorem and we may work in a neighborhood of one point p. In the language of Definition 4.1, we must show that the only possible point modifications of a 2-dimensional manifold by a 2-dimensional manifold are the finite iterations of quadratic transformations.

Let $A = \pi^{-1}(p)$. A is necessarily connected since π is proper. If A is one point, π is biholomorphic and no quadratic transformations are required. Assuming then that A is not a point, we shall first show that π may be factored through a quadratic transformation. Let (z_1, z_2) be local coordinates on M with $p = (0,0)$. In some neighborhood of A, π may then be given by a pair of functions $z_1 = F_1(s)$, $z_2 = F_2(s)$. We wish to now show that as $s \to q$, $s \notin A$ and $q \in A$, $(F_1(s), F_2(s))$, as homogeneous coordinates for a point in P^1, has a limit in P^1. Choose a local coordinate system (x,y) on M' with $q = (0,0)$. π is locally given by $z_1 = F_1(x,y)$, $z_2 = F_2(x,y)$ with $F_1(0,0) = F_2(0,0) = 0$. Let $t(x,y)$ be the greatest common divisor of F_1 and F_2; $F_1 = t(x,y)\, g_1(x,y)$, $F_2 = t(x,y)\, g_2(x,y)$. $(F_1, F_2) = (g_1, g_2)$ and we shall first show that $g_1(0,0) = g_2(0,0) = 0$ is impossible. Suppose then that $g_1(0,0) = g_2(0,0) = 0$. g_1 and g_2 are relatively prime so that for complex numbers $r \neq r'$, $rg_1 + g_2$ and $r'g_1 + g_2$ are relatively prime. Let h generate the ideal of A near (0,0). Any prime factor of h can thus divide $rg_1 + g_2$ for at most one value of r. h has only a finite number of prime factors so r may be chosen so that h and $rg_1 + g_2$ are relatively prime. By G & R, VIII. B.3, $V(rg_1+g_2)$, the zero set of $rg_1 + g_2$, and $A = V(h)$ intersect in a set of codimension 2, i.e. just (0,0).

$$\pi(V(rg_1+g_2)) \subset \{rz_1+z_2 = 0\}. \quad \pi(V(t)) = 0 \text{ so } V(t) \subset A.$$

Let $\rho : S \to V(rg_1+g_2)$ be a resolution of $V(rg_1+g_2)$ near (0,0). $\pi \circ \rho : S \to \{rz_1+z_2 = 0\}$ is a map between 1-dimensional manifolds which is non-constant on each component of S (i.e. on the resolution of each irreducible component of $V(rg_1+g_2)$) with $\pi \circ \rho(\rho^{-1}(0,0)) = (0,0)$. But $\pi \circ \rho$ is one-to-one off $\rho^{-1}(0,0)$ since both ρ and π are one-to-one there. Hence S has one component and $\pi \circ \rho$ is biholomorphic. Hence $\rho \circ (\pi \circ \rho)^{-1} = \pi^{-1} : \{rz_1+z_2 = 0\} \to V(rg_1+g_2)$ is holomorphic. Let $z = z_1$ be a coordinate for the complex line $rz_1 + z_2 = 0$. π^{-1} may then be given by some holomorphic functions $x = u(z)$ and $y = v(z)$ with $u(0) = v(0) = 0$. $\{rz_1+z_2 = 0\} \to \{z_2 = 0\}$ given by projection is a biholomorphic map which may be written $z_1 = F_1(u(z), v(z)) = F_1(z) = t(u(z), v(z))\, g_1(u(z), v(z))$. $F_1'(0) \neq 0$, so differentiating via the chain rule

$$\left[\frac{\partial t}{\partial x} u'(0) + \frac{\partial t}{\partial y} v'(0)\right] g_1(0,0) + t(0,0)\left[\frac{\partial g_1}{\partial x} u'(0) + \frac{\partial g_1}{\partial y} v'(0)\right]$$

$$= t(0,0)\left[\frac{\partial g_1}{\partial x} u'(0) + \frac{\partial g_1}{\partial y} v'(0)\right] \neq 0.$$

Hence $t(0,0) \neq 0$ and t is a unit. But F_1 and F_2 vanish on A, a set of codimension 1 and hence must have a common non-trivial factor by G & R, VIII. B. This is the desired contradiction. Hence either $g_1(0,0) \neq 0$ or $g_2(0,0) \neq 0$.

Hence (F_1,F_2) gives a well defined point in P^1. Let $(z_1,z_2) = (\xi\zeta, \zeta) = (\zeta', \xi'\zeta')$ be a quadratic transformation $\pi' : \tilde{M}' \to M$.

Suppose $g_2(0,0) \neq 0$. In the first coordinate system $(\xi, \zeta) = (z_1/z_2, z_2)$ and the map $\phi : (x,y) \to (\xi, \zeta) = (\frac{F_1(x,y)}{F_2(x,y)}, F_2(x,y))$ $= (\frac{g_1(x,y)}{g_2(x,y)}, t(x,y)\, g_2(x,y))$ is a holomorphic map. Hence $(\pi')^{-1} \circ \pi$, defined on $M' - A$, extends to a necessarily unique and proper holomorphic map $\phi' : M' \to \tilde{M}'$.

But ϕ' satisfies the hypotheses of the theorem. Hence we may repeat all the previous arguments and factor ϕ' through $\phi'' : M' \to \tilde{M}''$, where $\tilde{M}''$ is obtained from $\tilde{M}'$ by a quadratic transformation at some point on $\tilde{M}'$ in $(\pi')^{-1}(p)$. Continue to repeat the arguments. $\pi^{-1}(S)$ has only a finite number of 1-dimensional irreducible components. $\phi^{(\nu)}(\pi^{-1}(S))$ can have at most as many 1-dimensional irreducible components as does $\pi^{-1}(S)$. Hence after a finite number of steps we cannot repeat the factorization because $\phi^{(\nu)} : M' \to \tilde{M}^{(\nu)}$ has discrete fibres. Then, as shown above, $\phi^{(\nu)}$ is biholomorphic. $\tilde{M}^{(\nu)}$ is obtained from M by a finite number of iterated quadratic transformations as required.∎

DEFINITION 5.2. *A 1-dimensional analytic subset* A *in a 2-dimensional complex manifold is exceptional of the first kind if there is a proper holomorphic map* $\phi : M \to Y$ *with* Y *a manifold such that* $\Phi(A)$ *is a point* p *and* $\Phi : M - A \to Y - p$ *is biholomorphic. If* A *is irreducible, it is called an exceptional curve of the first kind.*

Proposition 4.5 insures that exceptional sets of the first kind are exceptional as in Definition 4.2.

COROLLARY 5.8. *A 1-dimensional analytic subset* A *in a 2-dimensional complex manifold* M *is exceptional of the first kind if and only if it is compact, connected and has a weighted graph which upon successively collapsing vertices with genus* 0 *and weight* -1 *becomes the empty graph.*

Proof: If A is exceptional of the first kind, Theorem 5.7 says that A is the result of an iterated sequence of quadratic transformations. Thus the components of A will be non-singular and will intersect each other transversely at no more than one point. Hence A will have a weighted graph and at least one vertex will be a curve of genus 0 and self-intersection number -1. Let $\pi: M \to M'$ collapse this curve. M' is a new manifold with $A' = \pi(A)$ as an exceptional set. $\Phi: M \to Y$ factors as $\Phi = \Phi' \circ \pi$, $\Phi': M' \to Y$. Thus A' is also exceptional of the first kind. A' has one less vertex in its weighted graph. After a finite number of such collapses, the graph of $A^{(n)}$ becomes the empty set.

Conversely if $\pi_1: M \to M'$, $\pi_2: M' \to M''$, ... reduces the weighted graph of A to the empty set, the composition of the π's exhibits A as an exceptional set of the first kind.∎

DEFINITION 5.3. *A resolution* $\pi: M \to V$ *of the singularities of* V *is a minimal resolution if for any other resolution* $\pi': M' \to V$ *there is a unique holomorphic map* $\rho: M' \to M$ *such that* $\pi' = \pi \circ \rho$.

Since the regular points of V are dense, any holomorphic ρ is necessarily unique. The following very general argument shows that if a minimal resolution exists then it is unique. Suppose

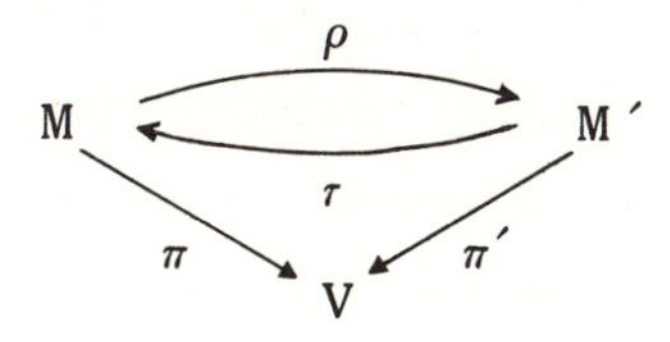

with $\pi: M \to V$ and $\pi': M \to V$ minimal resolutions. ρ and τ exist. $\tau \circ \rho$ is the identity map since $\pi: M \to V$ factors through $\pi: M \to V$ in a unique manner. Similarly $\rho \circ \tau$ is the identity and M and M' are biholomorphic. We shall thus speak of the minimal resolution, provided that it exists.

THEOREM 5.9. *Let* V *be a 2-dimensional normal space and let* $\{p_i\}$ *be the singularities of* V. *Let* $\tilde{\pi}: \tilde{M} \to V$ *be any resolution of* V.

The minimal resolution of V *may be obtained from* $\tilde{M}$ *by successively collapsing all exceptional curves of the first kind which lie above the* p_i.

Proof: Let $\pi: M \to V$ be the resolution of the theorem and let $\pi': M' \to V$ be another resolution. Since the theorem is local (since the singular points of V are isolated) it suffices to assume that V has only one singular point p and that the regular points of V are connected.

By construction, $\pi^{-1}(p)$ has no irreducible components which are non-singular curves of genus 0 with self-intersection number -1. Let $G = \{(q_1,q_2) \in M \times M' | \pi(q_1) = \pi'(q_2)\}$. G is a subvariety in $M \times M'$. Let ρ_1 and ρ_2 be the projection maps of $M \times M'$ onto M and M' respectively. Let $R = \{(q_1,q_2) \in G | \pi(q_1) \neq p\}$. R is connected since it is biholomorphic with the regular points of V. Let H be the closure of that connected component of the regular points of G which contains R. H is a subvariety in $M \times M'$ by III.C. 19 of G & R (although H need not be locally irreducible) of pure dimension 2. $S = H \cap \{(\pi \circ \rho_1)^{-1}(p)\}$ is a proper subvariety of H. Since the regular points of H are connected, S is of dimension at most 1. Let $\pi'': M'' \to H$ be a resolution of the singularities of H. Then $\theta_1 = \rho_1 \circ \pi'': M'' \to M$ and $\theta_2 = \rho_2 \circ \pi'': M'' \to M'$ are proper holomorphic maps between 2-dimensional manifolds which are biholomorphic off $\pi^{-1}(p)$ and $(\pi')^{-1}(p)$ respectively. Hence by Theorem 5.7, θ_1 and θ_2 are obtained by a finite number of iterated quadratic transformations. Also, since we have not yet made use of the fact that $\pi^{-1}(p)$ contains no exceptional curves of the first kind, we have proved

THEOREM 5.10. *If* $\pi: M \to V$ *and* $\pi': M' \to V$ *are resolutions of the normal 2-dimensional singularity* p, *then there exist a resolution* $\pi'': M'' \to V$ *and factorizations* $\theta_1: M'' \to M$ *and* $\theta_2: M'' \to M'$ *such that* $\pi'' = \pi \circ \theta_1 = \pi' \circ \theta_2$. θ_1 *and* θ_2 *are iterated quadratic transformations.*

In other words, starting from the resolution $\pi: M \to V$ with $\pi^{-1}(p) = A = \cup A_i$, the A_i irreducible, we may reach any other resolution $\pi': M' \to V$ with $(\pi')^{-1}(p) = A' = \cup A'_j$, the A'_j irreducible, as follows.

First perform a finite number of quadratic transformations in M, above p. This introduces new irreducible curves $\{B_k\}$. We then successively collapse various exceptional curves of the first kind until reaching M'. To prove Theorem 5.9, it thus suffices to show that if $A_i \subset \pi^{-1}(p)$ is not an exceptional curve of the first kind, then A_i cannot become an exceptional curve of the first kind via quadratic transformations on M. For then the first curve to be collapsed in going from M'' to M' must be a B_k. $\theta_1(B_k)$ and $\theta_2(B_k)$ are points. Let M''' be the manifold obtained by collapsing this B_k. By Theorem 5.7, M''' is obtained from M by iterated quadratic transformations and also from M' be iterated quadratic transformations. Also, the exceptional set in M''' has one less irreducible component than the exceptional set in M''. Repeat the argument. After a finite number of steps, we obtain $M^{(n)}$, equal to either M' or M and obtained (respectively) from M or M' by quadratic transformations. $M^{(n)} = M'$ must always occur for otherwise $M^{(n)} = M$ and $M^{(n)}$ is obtained from M' by a non-empty set of quadratic transformations. Then A, the exceptional set in $M^{(n)} = M$ contains an exceptional curve of the first kind, contrary to construction.

We thus must show that if A_i is not exceptional of the first kind, quadratic transformations cannot make A_i exceptional of the kind.

LEMMA 5.11. *Let* $A = \cup A_i$ *be the decomposition into irreducible components of an exceptional set* A *in the* 2*-dimensional manifold* M. *Let* B *be the union of a subset of the* $\{A_i\}$. *Then* B *is exceptional in* M.

Proof: Perform quadratic transformations $\tilde{\pi}\colon \tilde{M} \to M$ until $\tilde{\pi}^{-1}(A)$ has non-singular irreducible components which intersect transversely. Apply Theorem 4.9. Since any submatrix of a negative definite matrix whose entries are positioned symmetrically with respect to the diagonal is negative definite, B is exceptional.∎

So now suppose that A_i becomes exceptional of the first kind after a sequence of quadratic transformations. It suffices to consider only the shortest initial segment of the sequence which makes A_i into $\tilde{A}_1$, an exceptional curve of the first kind. Let B_e be the new curve introduced by the last quadratic transformation of the segment. Then $B_e \cdot B_e = -1$, $\tilde{A}_i \cdot \tilde{A}_i = -1$ and $B_e \cap \tilde{A}_i \neq \emptyset$ (for otherwise the last quadratic transformation does not affect A_i). Let us contradict Lemma 5.11 by showing that $\tilde{A}_i \cup B_e$ is not exceptional.

Let $\{q_t\} = \tilde{A}_i \cap B_e$. Perform quadratic transformations at the $\{q_t\}$ until $\tilde{A}_i$ and B_e do not intersect. Let n be the total number of quadratic transformations required to just separate B_e and $\tilde{A}_i$. We get a weighted graph.

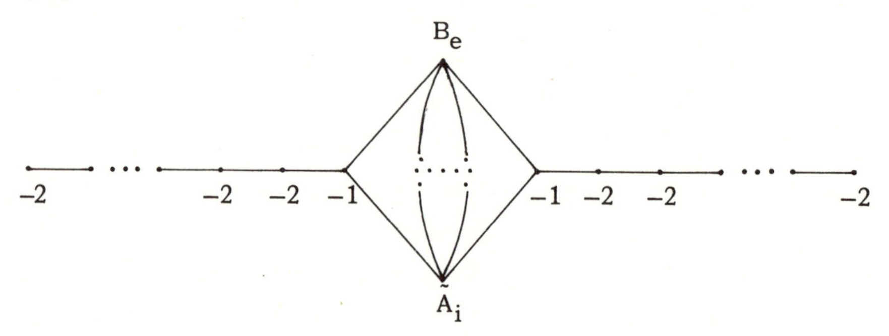

with $\tilde{A}_i \cdot \tilde{A}_i = B_e \cdot B_e = -1 - n$.

$$(5.11)\quad \begin{pmatrix}
-1-n & 0 & 1 & 0\ldots\ldots\ldots 0 & & 1 & 0\ldots\ldots\ldots 0\ldots\ldots\ldots \\
0 & -1-n & 1 & 0\ldots\ldots\ldots 0 & & 1 & 0\ldots\ldots\ldots 0\ldots\ldots\ldots \\
1 & 1 & -1 & 1 & & & & \\
0 & 0 & 1 & -2 & 1 & & & \\
0 & 0 & & 1 & -2 & 1 & 0 & \\
\vdots & & & & \ddots & & & \\
0 & & & & 1 & -2 & 0 & \\
1 & 1 & & & 0 & 0 & -1 & 1 \\
0 & 0 & & & & & 1 & -2 \\
\vdots & \vdots & & & & & & \\
0 & 0 & & & & & & \\
\vdots & \vdots & & & & & & \vdots
\end{pmatrix}$$

Using Lemma 4.2 and collapsing algebraically the new curves introduced by performing quadratic transformations at the $\{q_t\}$, we see that (5.11) is negative definite if and only if $\begin{pmatrix} -1 & n \\ n & -1 \end{pmatrix}$ is negative definite. But $\begin{pmatrix} -1 & n \\ n & -1 \end{pmatrix}$ is negative definite if and only if $n = 0$, i.e. if and only if $\tilde{A}_i$ and B_e do not intersect at all.∎

THEOREM 5.12. *Let* p *be a normal 2-dimensional singularity and* V *a neighborhood of* p *having* p *as its only singular point. Then there is a unique minimal resolution* $\pi\colon M \to V$ *among all resolutions satisfying conditions* (i), (ii) *and* (iii) *below. Let* $\pi^{-1}(p) = A = \cup A_i$ *be the decomposition of* $\pi^{-1}(p)$ *into irreducible components.*

(i) *Each* A_i *is non-singular*

(ii) A_i *and* A_j, $i \neq j$, *intersect transversely wherever they intersect*

(iii) *No three distinct* A_i *meet.*

Proof: Let $\tilde{\pi}\colon \tilde{M} \to V$ be the minimal resolution of p from Theorem 5.9. Let q be a singular point of $\tilde{\pi}^{-1}(p)$. If q is either (i′) a singular point of an $\tilde{A}_i$, (ii′) a non-transverse intersection point of $\tilde{A}_i$ and $\tilde{A}_j$. $i \neq j$ or (iii′) a point of three distinct $\tilde{A}_i$, perform a quadratic transformation at q to get a resolution $\tilde{\pi}'\colon \tilde{M}' \to V$. If $(\tilde{\pi}')^{-1}(p)$ has a point q' satisfying (i′), (ii′) or (iii′) perform a quadratic transformation at q'. After a finite number of such quadratic transformations we obtain a resolution $\pi\colon M \to V$ satisfying (i), (ii) and (iii). π is in fact the desired minimal resolution for suppose $\pi'\colon M' \to V$ is another resolution having properties (i), (ii) and (iii). $\pi' = \tilde{\pi} \circ \theta_0$ where θ_0 consists of finite iterations of quadratic transformations at the points $r_1, \ldots, r_s$. Since π' has properties (i), (ii) and (iii), $\{r_i\}$ must contain all points $\{q_j\}$ in $\tilde{A}$ with properties (i′), (ii′) or (iii′). Let $\tilde{\pi}_1$ be obtained from $\tilde{\pi}$ by quadratic transformations at the $\{q_j\}$.

$\pi' = \tilde{\pi}_1 \circ \theta_1$ and θ_1 is the finite iteration of quadratic transformations at the points $\{r'_i\}$. Again since π' has properties (i), (ii) and (iii), $\{r'_i\}$ must contain all points $\{q'_j\} \in \tilde{A}_1$ with properties (i′), (ii′), or (iii′) and may contain additional points. Let $\tilde{\pi}_2$ be obtained from $\tilde{\pi}_1$ by quadratic transformations at the $\{q'_j\}$. Again $\pi' = \tilde{\pi}_2 \circ \theta_2$. After a finite number of such factorizations, $\tilde{\pi}_n = \pi$, as desired.

Uniqueness follows as in the case of the usual minimal resolution. ▮

CHAPTER VI

EQUIVALENCE OF EMBEDDINGS

So far, in considering our general problem of classifying normal singularities of surfaces, we know that each normal singularity has a unique minimal resolution, Theorem 5.9, and also a unique minimal resolution in which the exceptional set is "nice," Theorem 5.12. Also, if two resolutions are isomorphic in a neighborhood of the exceptional set, then by Theorem 3.13, the singularities themselves have biholomorphic neighborhoods. Thus, to describe all singularities it suffices to describe all minimal resolutions (as in either Theorem 5.9 or Theorem 5.12). It will be convenient for us to assume that A has the properties (i), (ii), and (iii) of Theorem 5.12. The question considered in this chapter is: In how many ways can A be embedded in 2-dimensional complex manifolds with given weights? Our answer is not quite complete but, roughly speaking, we shall show that only a finite number of conditions must be satisfied in order for two embeddings to be equivalent.

It should be pointed out that it is possible for two different embeddings of A to give the same singularity. Let, for example, A be a given Riemann surface of genus 1. Let q and r be different points on A. Let N be the bundle corresponding to the divisor $-q$ and let $\tilde{N}$ correspond to the divisor $-r$. Let $\phi : A \to N$ and $\tilde{\phi} : A \to \tilde{N}$ be the embeddings of A onto the 0-sections of the bundles. ϕ and $\tilde{\phi}$ are different embeddings, for consider any biholomorphic map $\psi : U \to \tilde{U}$, where U and $\tilde{U}$ are neighborhoods of A and $\tilde{A}$ respectively, such that $\psi \circ \phi = \tilde{\phi}$. ψ induces an isomorphism of the normal bundles of the embeddings ϕ and $\tilde{\phi}$. But N and $\tilde{N}$ are distinct bundles since their difference corresponds to the bundle r−q. This is a nontrivial bundle since there are no elliptic functions with a single simple pole. Thus ψ cannot exist. But in order for the total spaces of N and $\tilde{N}$ to be biholomorphic, we do not need that $\psi \circ \phi = \tilde{\phi}$. There is, in fact, an automorphism $\theta: A \to A$ such that $\theta(q) = r$. θ is induced by a translation on C, the universal covering space of A. θ induces a biholomorphic map between N and $\tilde{N}$. Thus N and $\tilde{N}$ determine the same singularity.

We shall resolve the above difficulty in the next section, but we now return to the problems of embeddings. We shall proceed in two steps. The first is to establish conditions for a formal equivalence; the second is to show that a formal equivalence implies an actual equivalence.

VIa. Spaces with Nilpotents

Let us start with the simplest possible embedding, the origin 0 embedded in C. Let $\mathcal{O}$ be the sheaf of germs of holomorphic functions on C. The embedding of 0 in C is completely described by the pair $(0, \mathcal{O}_0)$ with $\mathcal{O}_0$ being of course the ring of convergent power series in one variable. Namely, if $p \in M$, an analytic space, 0 and p have isomorphic neighborhoods if and only if $\mathcal{O}_0$ and $\mathcal{O}_p$ are isomorphic as rings. We may "approximate" $\mathcal{O}_0$, and thereby the embedding of 0 in C, as follows. Let (z^2) be the ideal sheaf generated by z^2. $\mathcal{O}/(z^2)$ is supported at the origin. $(\mathcal{O}/(z^2))_0 \approx C^2$ and might suggestively be written $(\mathcal{O}/(z^2))_0 \approx C \oplus C\,z$. The "function" z is nilpotent in $\mathcal{O}/(z^2)$, i.e., $z \neq 0$ in $\mathcal{O}/(z^2)$ but $z^2 = 0$ in $\mathcal{O}/(z^2)$. Looking successively at $\mathcal{O}/(z)$, $\mathcal{O}/(z^2)$, $\mathcal{O}/(z^3), \ldots$ and the natural projection maps $\mathcal{O}/(z^\nu) \to \mathcal{O}/(z^{\nu-1})$ we have

$$\cdots \to C \oplus C\,z \oplus C\,z^2 \to C \oplus C\,z \to C.$$

Passing to the inverse limit, we get the ring of formal power series in one variable, which "formally" describes the embedding of 0 in C.

The situation in higher dimensions is much more complicated but is roughly the same in character. It will be very convenient to generalize the notion of analytic spaces so as to include nilpotent elements such as z above. We will eventually use these spaces to more fully describe embeddings.

Recall, Chapter V of G & R, that an analytic space V is an object that may be defined as a ringed space $(V, {}_V\mathcal{O})$ with V an underlying Hausdorff space and ${}_V\mathcal{O}$ a sheaf of rings over V. ${}_V\mathcal{O}$ is often called the structure sheaf. Locally V may be embedded as a subvariety in a polydisc $\Delta \subset \mathbf{C}^n$. Let $\mathcal{O}$ be the sheaf of germs of holomorphic functions on Δ. In Δ, if $\mathcal{I}$ is the ideal sheaf of V, ${}_V\mathcal{O} \approx (\mathcal{O}/\mathcal{I} \mid V)$. More generally, if $\mathcal{F}$ is any coherent sheaf of ideals in $\mathcal{O}$, we may let V be the locus of $\mathcal{F}$ and ${}_V\mathcal{O} = (\mathcal{O}/\mathcal{F} \mid V)$. Recall that if $\theta : R \to S$ is a continuous map and $\mathcal{H}$ is a sheaf over R, then $\theta_* (\mathcal{H})$ is a sheaf over S given by Definition 5.1. Let $\theta^* : \Gamma(U, \theta_*(\mathcal{H})) \to \Gamma(\theta^{-1}(U), \mathcal{H})$ denote the isomorphism which defines the presheaf of $\theta_*(\mathcal{H})$. We can now give the following formal definition.

DEFINITION 6.1. *An analytic space with nilpotents, also called a nonreduced space, is a pair* $(V, {}_V\mathcal{O})$ *where* V *is a Hausdorff topological space and* ${}_V\mathcal{O}$*, the structure sheaf, is a sheaf of* $\mathbf{C}$-algebras *over* V *such that every* $p \in V$ *has a neighborhood* U *as follows. There is a homeomorphism* $\iota : U \to Y$*, where* Y *is the locus of a coherent sheaf of ideals* $\mathcal{F}$ *in a polydisc in some* $\mathbf{C}^n$ *and there is an isomorphism* $\phi : (\mathcal{O}/\mathcal{F} \mid Y) \to \iota_*({}_V\mathcal{O} \mid U)$ *of sheaves of* $\mathbf{C}$-algebras *over* Y.

Two analytic spaces with nilpotents $(V, {}_V\mathcal{O})$ and $(W, {}_W\mathcal{O})$ are isomorphic via (θ, ψ) if there are a homeomorphism $\theta : V \to W$ and an isomorphism $\psi : {}_W\mathcal{O} \to \theta_*({}_V\mathcal{O})$ of sheaves of $\mathbf{C}$-algebras over W. Let $\theta^* : \theta_*({}_V\mathcal{O}) \to {}_V\mathcal{O}$ also denote the mapping of the total spaces induced by the mapping θ^* of the presheaves. Suppose that $(\omega, \phi) : (W, {}_W\mathcal{O}) \to (S, {}_S\mathcal{O})$ is another isomorphism.

$(\omega, \phi) \circ (\theta, \psi) = (\omega \circ \theta, ((\omega \circ \theta)^*)^{-1} \circ \theta^* \circ \psi \circ \omega^* \circ \phi)$: $(V, {}_V\mathcal{O}) \to (S, {}_S\mathcal{O})$ is the composite isomorphism.

$(\theta^{-1}, ((\theta^{-1})^*)^{-1} \circ \psi^{-1} \circ (\theta^*)^{-1}) : (W, {}_W\mathcal{O}) \to (V, {}_V\mathcal{O})$ is the inverse to (θ, ψ).

A sheaf $\mathcal{S}$ of ${}_V\mathcal{O}$ modules over V is coherent if in some neighborhood of each point $p \in V$, there is a resolution

$$ {}_V\mathcal{O}^r \to {}_V\mathcal{O}^s \to \mathcal{S} \to 0 . $$

If $\mathcal{G}$ is a coherent sheaf of ideals and $X = \operatorname{loc} \mathcal{G}$, $(X, {}_V\mathcal{O}/\mathcal{G})$ is a subspace with nilpotents of $(V, {}_V\mathcal{O})$.

Analytic spaces with nilpotents differ from ordinary, or reduced spaces, essentially only in that the structure sheaf may have nilpotent elements. From now on in this section, we shall omit the words "with nilpotents"; all analytic spaces in this section will be assumed to have nilpotent elements in the structure sheaf unless otherwise stated. We shall abuse our notation in the usual manner by frequently omitting reference to the structure sheaf when referring to an analytic space.

DEFINITION 6.2. *Let* $p \in V$, *an analytic space, and let* m *be the maximal ideal in* ${}_V\mathcal{O}_p$. *The tangential dimensional of* V *of* p, *denoted* $\operatorname{dimt}_p V$, *is the dimension of* m/m^2. *A neat embedding of* V *at* p *is an isomorphism* ϕ *of a neighborhood of* p *onto a substance (with nilpotents) of a domain in* C^k, $k = \operatorname{dimt}_p V$.

Locally, as in Definition 6.1, ${}_V\mathcal{O} \approx \mathcal{O}/\mathcal{F}$. Let $\pi : \mathcal{O}_{\iota(p)} \to {}_V\mathcal{O}_p$ be the projection map and let M be the maximal ideal of $\mathcal{O}_{\iota(p)}$. Then $\pi(M) = m$ and $\pi(M^2) = m^2$. Hence there is an induced surjective map $\pi_*: M/M^2 \to m/m^2$ and $k \le n$.

THEOREM 6.1. *Let* V *be a subspace of a polydisc* Δ *in* C^n *and* $p \in V$. *Let* $k = \operatorname{dimt}_p V$. *Then there is a complex submanifold* R *of* C^n

of dimension k *such that* V *may be expressed as a subspace of* R *in a neighborhood of* p.

Proof: The proof is by induction on n. $n = 0$ is trivial since $\mathbb{C}$ is a field.

Suppose $k < n$. Then π_* has a nontrivial kernel, i.e., letting $(z_1, \ldots, z_n)$ be coordinates in $\mathbb{C}^n$ with $p = (0, \ldots, 0)$ there are constants $(a_1, \ldots, a_n) \neq (0, \ldots, 0)$ such that $\Sigma (a_i z_i)_o \equiv \Sigma (f_j g_j)_o$, mod $\mathcal{F}_o$; $(f_j)_o$ and $(g_j)_o \in M$. By the implicit mapping theorem, we may choose new coordinates $(\zeta_1, \ldots, \zeta_n)$, with $\zeta_1 = \Sigma\, a_i z_i - \Sigma\, f_j g_j$, in some neighborhood of the origin. Let $S = \{\zeta_1 = 0\}$, a subspace of dimension $n-1$. $\mathcal{O}/\mathcal{F} \approx (\mathcal{O}/(\zeta_1)) / (\mathcal{F}/(\zeta_1))$ since ζ_1 is a section of $\mathcal{F}$. $\mathcal{O}/(\zeta_1)$ is isomorphic to the structure sheaf for $\mathbb{C}^{n-1}$ so we may apply the induction hypothesis.∎

THEOREM 6.2. *Let* $V \ni p$ *and* $W \ni q$ *be analytic spaces. Let* $R \subset \Delta \subset \mathbb{C}^k$ *and* $S \subset \Delta' \subset \mathbb{C}^{k'}$ *be the images of neat embeddings of* V *and* W *near* p *and* q *respectively with* p *and* q *corresponding to the origins. Let* $\mathcal{F}$ *and* $\mathcal{G}$ *be such that* ${}_R\mathcal{O} \approx \mathcal{O}/\mathcal{F}$ *and* ${}_S\mathcal{O} \approx \mathcal{O}/\mathcal{G}$ *and let* $\pi_1 : \mathcal{O} \to {}_R\mathcal{O}$ *and* $\pi_2 : \mathcal{O} \to {}_S\mathcal{O}$ *denote the projection maps. Suppose that* $\theta : R \to S$ *and* $\psi : {}_S\mathcal{O} \to \theta_*({}_R\mathcal{O})$ *define an isomorphism with* $\theta(p) = q$. *Let* $(z_1, \ldots, z_k)$ *be ambient coordinate functions near* p. *Let* $(\zeta_1, \ldots, \zeta_k)$ *be any ambient functions near* q *such that* $\pi_1(z_i) = \theta^* \circ \psi \circ \pi_2(\zeta_i)$. *Then* $(\zeta_1, \ldots, \zeta_k)$ *are ambient coordinate functions near* q. *Let* $\Psi(z_i) = \zeta_i$ *be a map of the ambient spaces. Then near the origin,* $\Psi^*(\mathcal{G}) = \mathcal{F}$ *and* Ψ *induces* θ *and* ψ.

Theorem 6.2 corresponds to Theorem V.B.16 of G & R and simply says that any isomorphism of neatly embedded analytic spaces extends locally to a biholomorphic map of the ambient spaces.

Proof of Theorem 6.2. Let M be the ambient maximal ideal at the origin. $(\zeta_1, \ldots, \zeta_k)$ are local coordinates since any set of functions projecting onto a basis of M/M^2 in a one-to-one manner is a set of local coordinates, and conversely.

Elements of ${}_R\mathcal{O}_r$, $r \in R$, can be evaluated as functions at r by looking at their images in ${}_V\mathcal{O}_r/m$, where m is the maximal ideal. ${}_V\mathcal{O}_r/m$ is canonically isomorphic to $\mathbf{C}$ via the $\mathbf{C}$-algebra structure. Near the origin, $(z_1, \ldots, z_k)$ separates points, as does $(\zeta_1, \ldots, \zeta_k)$. Hence Ψ induces θ.

Ψ was chosen so that $\Psi^*(\zeta_i)_o + \mathcal{F}_o = (z_i)_o + \mathcal{F}_o = (\theta^* \psi\, \pi_2(\zeta_i))_o$. $\Psi^*(g(\zeta))_o + \mathcal{F}_o = g(z)_o + \mathcal{F}_o$. We need that

$$(6.1) \qquad g(z)_o + \mathcal{F}_o = (\theta^* \psi\, \pi_2(g(\zeta))_o$$

for all convergent power series g. Both Ψ^* and $\theta^* \circ \psi \circ \pi_2$ are $\mathbf{C}$-algebra homomorphisms, so (6.1) holds for g a polynomial. Omitting the stalk subscript, suppose $\Psi^*(g) + \mathcal{F} \neq \theta^* \circ \psi \circ \pi_2(g)$ for some g. Then, letting m be the maximal ideal in ${}_R\mathcal{O}_o$, $\Psi^*(g) + \mathcal{F} - \theta^* \circ \psi \circ \pi_2(g) \in m^\nu - m^{\nu+1}$ for some ν. Let h be the polynomial representing the power series terms of g up to and including all homogeneous terms of degree ν. $\Psi^*(h) + \mathcal{F} = \theta^* \circ \psi \circ \pi_2(h)$ so $\Psi^*(g-h) + \mathcal{F} - \theta^* \circ \psi \circ \pi_2(g-h) \in m^\nu - m^{\nu=1}$. But $g - h \in M^{\nu+1}$. Hence $\Psi^*(g-h) + \mathcal{F} \in m^{\nu+1}$ and $\theta^* \circ \psi \circ \pi_2(g-h) \in m^{\nu+1}$, a contradiction. Thus (6.1) holds for all convergent power series. Since $\theta^* \circ \psi$ is an isomorphism, $\mathcal{G}_o = \ker \theta^* \circ \psi \circ \pi_2 = \ker (\Psi^* + \mathcal{F}_o)$. Ψ^* is an isomorphism on the ambient functions, so $\Psi^*(\mathcal{G}_o) = \mathcal{F}_o$. Since $\mathcal{F}$ and $\mathcal{G}$ are coherent sheaves, $\Psi^*(\mathcal{G}) = \mathcal{F}$ in some neighborhood of the origin. Hence Ψ induces ψ and the proof is complete. ∎

We shall now set up still more machinery.

Let A be a reduced analytic subset of the manifold M. Let m be a coherent ideal sheaf on M with A as its locus. Let Ω be the sheaf of germs of holomorphic 1-forms on M. Let $\Omega' \subset \Omega$ be the subsheaf of Ω

generated by $m\Omega$ and df for $f \in m$, i.e., $\omega_x \in \Omega'_x$ if

$\omega_x = \Sigma h_i \theta_i + \Sigma g_j df_j$ with $\theta_i \in \Omega_x$, $g_j \in \mathcal{O}_x$ and $h_i, f_j \in m_x$. We call ${}_m\Omega = \Omega/\Omega'$ the sheaf of germs of holomorphic 1-forms on the analytic space $(A, \mathcal{O}/m) = A(m)$. Let ${}_m\mathcal{O}$ denote $\mathcal{O}/m$. Since $m\,\Omega \subset \Omega'$, ${}_m\Omega$ is a sheaf of ${}_m\mathcal{O}$-modules. We call ${}_m\Theta = \mathcal{H}om({}_m\Omega, {}_m\mathcal{O})$ the tangent sheaf to A (m).

Suppose now that $m \supset n$, another coherent ideal sheaf with $A = \text{loc } n$. Suppose also that $m^2 \subset n$. We have the following exact sequences of sheaves of abelian groups.

$$0 \to n \to m \to m/n \to 0$$

$$0 \to m/n \to {}_n\mathcal{O} \to {}_m\mathcal{O} \to 0$$

The projection map ${}_n\mathcal{O} \to {}_m\mathcal{O}$ gives every ${}_m\mathcal{O}$-module an ${}_n\mathcal{O}$-module structure.

PROPOSITION 6.3. *${}_m\Omega$ is a coherent sheaf of ${}_n\mathcal{O}$-modules.*

Proof: We essentially repeat the proof that appeared in the proof of Theorem 3.7. Locally $A \subset \Delta \subset C^r$ and ${}_m\mathcal{O} = \mathcal{O}/m$. Thus any ${}_m\mathcal{O}$-module has an induced $\mathcal{O}$-module structure. Ω is a free sheaf in Δ. Let $f_1, \ldots, f_t$ generate m near 0 in Δ. $\Omega' \subset \Omega$ is the image of the sheaf map $\gamma: \mathcal{O}^{tr+t} \to \Omega$ given by $\gamma(1,0,\ldots,0) = f_1 dz_1$, $\gamma(0,1,0,\ldots,0) = f_1 dz_2, \ldots, \gamma(0,\ldots,0,1,0,\ldots,0) = f_t dz_r$, $\gamma(0,\ldots,0,1,0,\ldots) = df_1, \ldots, \gamma(0,\ldots,0,1) = df_t$. Hence Ω' is a coherent sheaf of $\mathcal{O}$-modules and then so must ${}_m\Omega$ be a coherent sheaf of $\mathcal{O}$-modules. Let

$$\mathcal{O}^u \overset{\alpha}{\to} \mathcal{O}^v \overset{\beta}{\to} {}_m\Omega \to 0$$

resolve ${}_m\Omega$ locally. Taking quotients with respect to n, we get a resolution

$${}_n\mathcal{O}^u \overset{\alpha *}{\to} {}_n\mathcal{O}^v \overset{\beta *}{\to} {}_m\Omega \to 0 . ∎$$

Proposition 3.8 goes through without change. Hence

$$ {}_m\Theta_x = \mathcal{H}om\,({}_m\Omega, {}_m\mathcal{O})_x = \mathrm{Hom}_{{}_m\mathcal{O}_x}({}_m\Omega_x, {}_m\mathcal{O}_x)\,. $$

We now define some sheaves which are in general sheaves of non-abelian groups. This presents no trouble with the definitions of the sheaf structure but does lead to difficulties in defining sheaf cohomology. We will briefly discuss cohomology in sheaves of non-abelian groups later in this section. However, except in the proof of Proposition 6.21, all the sheaves that appear in the proofs of the theorems are shown to be actually abelian.

If $\alpha_x: {}_n\mathcal{O}_x \to {}_n\mathcal{O}_x$ is a $\mathbb{C}$-algebra automorphism such that $\alpha_x(m/n)_x \subset (m/n)_x$, then α_x induces a map $\tilde{\alpha}_x: {}_m\mathcal{O}_x \to {}_m\mathcal{O}_x$.

Let $\mathcal{A}ut\,(n:m)$ be the sheaf of automorphisms, with stalkwise multiplication given by composition, defined by the following complete presheaf. The maps on the underlying topological spaces are the identity; we shall omit references to the direct image sheaves. If U is open in A, $\Gamma(U, \mathcal{A}ut\,(n:m)) = \{$isomorphisms $\alpha:(U, {}_n\mathcal{O}\,|\,U) \to (U, {}_n\mathcal{O}\,|\,U)$ such that α is the identity map on U, $\alpha_x(m/n)_x = (m/n)_x$ and α_x induces the identity map $\tilde{\alpha}_x: {}_m\mathcal{O}_x \to {}_m\mathcal{O}_x$ for all $x \in U\}$. $\Gamma(U, \mathcal{A}ut\,(n:m))$ is a group under composition and satisfies the necessary compatibility conditions under restriction maps. We must verify that this is a complete presheaf, i.e., sections are determined by their local behavior and they patch together to give larger sections. But isomorphisms have been defined via sheaf maps and these are defined locally.

Let $\mathcal{A}ut\,(n,m)$ be the subsheaf of $\mathcal{A}ut\,(n:m)$ of maps such that $\alpha_x:(m/n)_x \to (m/n)_x$ is the identity map. For each open set U, $\Gamma(U, \mathcal{A}ut\,(n,m))$ is a normal subgroup of $\Gamma(U, \mathcal{A}ut(n:m))$ for let $\beta \in \Gamma(U, \mathcal{A}ut\,(n,m))$ and $\alpha \in \Gamma(U, \mathcal{A}ut\,(n:m))$. We must examine $(\alpha \circ \beta \circ \alpha^{-1})_x$ on $(m/n)_x$. But $(\alpha \circ \beta \circ \alpha^{-1})_x = (\alpha \circ \alpha^{-1})_x$ on $(m/n)_x$ since β is the identity map.

Hence $(\alpha \circ \beta \circ \alpha^{-1})$ is also the identity on $(m/n)_x$. Thus we may define the quotient sheaf of non-abelian groups, $\mathcal{A}n(n,m) = \mathcal{A}ut(n{:}m)/\mathcal{A}ut(n,m)$ by the (not necessarily complete) presheaf $\{\Gamma(U, \mathcal{A}ut(n{:}m))/\Gamma(U, \mathcal{A}ut(n,m))\}$.

$$(6.2) \qquad 1 \to \mathcal{A}ut(n,m) \to \mathcal{A}ut(n{:}m) \to \mathcal{A}n(n,m) \to 1$$

is an exact sheaf sequence.

It will be important to have a clearer representation of $\mathcal{A}ut(n{:}m)$ and $\mathcal{A}n(n,m)$. Let ${}_{n,m}\Theta = \mathcal{H}om_{{}_n\mathcal{O}}({}_m\Omega, m/n)$, where m/n has the natural ${}_n\mathcal{O}$-module structure. By Proposition 3.8, ${}_{n,m}\Theta_x = \mathrm{Hom}_{{}_n\mathcal{O}_x}({}_m\Omega_x, (m/n)_x)$.

PROPOSITION 6.4. *There is a sheaf isomorphism* $\lambda{:}{}_{n,m}\Theta \to \mathcal{A}ut(n,m)$ *given by* $(\lambda(\xi_x))(f_x) = f_x + \xi_x(df_x)$.

Proof: If locally $A \subset \Delta \subset \mathbf{C}^r$, the group homomorphism $d{:}\mathcal{O} \to \Omega$ induces a map $\tilde{d}{:}{}_m\mathcal{O} \to {}_m\Omega$ since $d(m) \subset \Omega'$. We also have the projection map $\pi{:}{}_n\mathcal{O} \to {}_m\mathcal{O}$. For $f_x \in {}_n\mathcal{O}_x$, $df_x = \tilde{d}(\pi f_x)$. We must now verify the proposition. First let us show that $\lambda(\xi_x) \in \mathcal{A}ut(n,m)_x$. $\lambda(\xi_x)$ is additive.

$$(\lambda(\xi_x))(f_x g_x) = f_x g_x + \xi_x(f_x dg_x + g_x df_x) =$$

$$= f_x g_x + f_x \xi_x(dg_x) + g_x \xi_x(df_x) .$$

On the other hand,

$$(\lambda(\xi_x))(f_x)(\lambda(\xi_x))(g_x) = [f_x + \xi_x(df_x)]\,[g_x + \xi_x(dg_x)] =$$

$$= f_x g_x + f_x \xi_x(dg_x) + g_x \xi_x(df_x) + \xi_x(df_x)\,\xi_x(dg_x) \quad .$$

But $m^2 \subset n$, so $\xi_x(df_x)\,\xi_x(dg_x)$ is trivial in ${}_n\mathcal{O}$. Hence $\lambda(\xi_x)$ is a ring homomorphism. If $f_x \in (m/n)_x$, $\pi(f_x) = 0$, so $\lambda(\xi_x)$ is the identity on m/n. Thus, $\lambda(-\xi_x)$ is the inverse to $\lambda(\xi_x)$. $\lambda(\xi_x)$ is $\mathbf{C}$-algebra automorphism. Since $\xi_x(df_x) \in (m/n)_x$, the induced map on ${}_m\mathcal{O}_x$ is the identity and $\lambda(\xi_x) \in \mathcal{A}ut(n,m)_x$ as desired.

We now must check that λ is an isomorphism. $\lambda(\eta_x + \xi_x)f_x =$
$= f_x + (\eta_x + \xi_x)(df_x) = f_x + \eta_x(df_x) + \xi_x(df_x)$.

$$\lambda(\eta_x) \circ \lambda(\xi_x)f_x = \lambda(\eta_x)\,[f_x + \xi_x(df_x)]$$
$$= f_x + \xi_x(df_x) + \eta_x(df_x) + \eta_x(\xi_x(df_x))\,.$$

Since $\xi_x(df_x) \in (m/n)_x$, $\eta_x(\xi_x(df_x)) = 0$ and λ_x is a group homomorphism. If $\lambda(\xi_x) = 1$, then $\xi_x(df_x) = 0$ for all f_x. But $\{df_x\}$ generate ${}_m\Omega$ as an ${}_n\mathcal{O}$-module so $\xi_x = 0$ and λ is an injection. We have only left to show that λ is surjective. Given $\alpha_x \in \mathcal{A}ut(n,m)_x$, we let $\tilde{\xi}_x\colon \mathcal{O} \to (m/n)_x$ be defined by $\tilde{\xi}_x(\tilde{f}_x) = \alpha_x(f_x) - f_x$ where f_x is the image in ${}_n\mathcal{O}_x$ of $\tilde{f}_x$. $\tilde{\xi}_x$ does indeed have its image in $(m/n)_x$ because α induces the identity on ${}_m\mathcal{O}$. We must verify that $\tilde{\xi}_x$ induces $\lambda^{-1}(\alpha_x) \in {}_{n,m}\Theta = \mathcal{H}om({}_m\Omega, m/n)$. We first show that $\tilde{\xi}_x$ acts well under Leibnitz's rule, i.e.,

$$\tilde{\xi}_x(\tilde{f}_x\tilde{g}_x) = \tilde{f}_x\,\tilde{\xi}_x(\tilde{g}_x) + \tilde{g}_x\,\tilde{\xi}_x(\tilde{f}_x)\,.$$

$$\tilde{\xi}_x(\tilde{f}_x\tilde{g}_x) = \alpha(f_xg_x) - f_xg_x = \alpha(f_xg_x) - f_x\alpha(g_x) + f_x\alpha(g_x) - f_xg_x$$
$$= \alpha(g_x)\,[\alpha(f_x) - f_x] + f_x[\alpha(g_x) - g_x]$$
$$= \alpha(g_x)\,\tilde{\xi}\,(\tilde{f}_x) + f_x\tilde{\xi}\,(\tilde{g}_x)\quad .$$

But $\alpha(g_x) - g_x \in (m/n)_x$ and $\tilde{\xi}\,(\tilde{f}_x) \in (m/n)_x$. Hence $\alpha(g_x)\,\tilde{\xi}\,(\tilde{f}_x) = g_x\,\tilde{\xi}(\tilde{f}_x)$ because $m^2 \subset n$. We now show that $\tilde{\xi}$ induces $\xi\colon \Omega \to m/n$ via $\xi(d\tilde{f}_x) = \tilde{\xi}(\tilde{f}_x)$. If (z_1, z_2) are local coordinates, we define ξ on all of Ω by

$$\xi(\Sigma\,\bar{g}_i\,dz_i)_x = \Sigma\,(\tilde{g}_i)_x\,\tilde{\xi}(z_i)_x\quad ,$$

which makes ξ an $\mathcal{O}$-module homomorphism but requires verification that the two definitions agree. But since $\tilde{\xi}$ acts well under Leibnitz's rule, the two definitions agree when $\tilde{f}_x$ is a polynomial in z_1 and z_2. Looking at representatives of the germs $\tilde{f}_x$ and $(\tilde{g}_i)_x$ in some fixed polydisc neighborhood, we see that $\tilde{\xi}$ is a continuous map since a is induced by a holomorphic map (Theorem 6.2). Polynomials are dense and both definitions of ξ are continuous so the two definitions agree. Our last task is to verify that $\xi(\Omega') = 0$. If $\tilde{f}_x \in m_x$ and $\omega_x \in \Omega_x$, $\xi(\omega_x) \in (m/n)_x$ and $\xi(\tilde{f}_x\omega_x) \in (m^2/n)_x = 0$. Also if $\tilde{f}_x \in m$,

$$\xi(d\tilde{f}_x) = \tilde{\xi}(\tilde{f}_x) = a(f_x) - f_x \ .$$

But a is the identity on m/n, so $\xi(d\tilde{f}_x) = 0$.∎

Proposition 6.4 shows in particular that $\mathcal{A}ut\,(n,m)$ is a sheaf of abelian groups.

We now specialize further and assume that A is a 1-dimensional reduced subspace of the 2-dimensional manifold M. We are not yet assuming that A is compact. Let $\{A_i\}$ be the irreducible components of A. We assume that the A_i are non-singular, cross transversely and no three meet at a point. Let $\mathcal{O}$ be the structure sheaf for M.

Let p and q be the ideal sheaves of A_1 and A_2, respectively. Let $m = p^s q^\nu$, $n = p^s q^{\nu+1}$ where s may be $0,1,2,\ldots$ but $\nu = 2,3,4,\ldots$. ${}_{n,m}\Theta = \mathcal{H}om_{{}_n\mathcal{O}}({}_m\Omega, m/n)$. $m/n \equiv 0$ off A_2 so ${}_{n,m}\Theta$ is supported on A_2. $m/n = p^s q^\nu / p^s q^{\nu+1}$. Since $qm = n$, m/n has a natural structure as an ${}_q\mathcal{O}$-module. Moreover, m/n is a locally free sheaf of ${}_q\mathcal{O}$-modules. If

$$a \in \mathcal{H}om_{{}_n\mathcal{O}}({}_m\Omega, m/n),\ \omega \in {}_m\Omega_x,\ f \in q_x,\ a(f_x\omega_x) = f_x a(\omega_x) = 0.$$

Hence a induces $\tilde{a}: {}_m\Omega/\Omega q \to m/n$, an ${}_q\mathcal{O}$-module homomorphism (as well as an ${}_n\mathcal{O}$-module homomorphism). Thus the natural map

$$\mathcal{H}om_{{}_n\mathcal{O}}({}_m\Omega/{}_m\Omega\,q, m/n) \to \mathcal{H}om_{{}_n\mathcal{O}}({}_m\Omega, m/n)$$

is an isomorphism. As ${}_n\mathcal{O}$-modules,

$$\mathcal{H}om_{{}_n\mathcal{O}}({}_m\Omega/{}_m\Omega\,q, m/n) \approx \mathcal{H}om_{{}_q\mathcal{O}}({}_m\Omega/{}_m\Omega\,q, m/n) \ .$$

As $\mathcal{O}$-modules, ${}_m\Omega/{}_m\Omega\,q \approx \Omega/\{$the $\mathcal{O}$-module generated by $q\Omega$ and $\Omega'\}$. Since $\nu \geq 2$, $\Omega' \subset q\,\Omega$ and ${}_m\Omega/{}_m\Omega\,q \approx \Omega/q\,\Omega$, which incidentally is independent of s and ν. Since Ω is a locally free sheaf of $\mathcal{O}$-modules, $\Omega/q\,\Omega$ is a locally free sheaf of ${}_q\mathcal{O}$-modules. m/n is also a locally free sheaf of ${}_q\mathcal{O}$-modules. Hence

$$\mathcal{H}om_{{}_q\mathcal{O}}({}_m\Omega/{}_m\Omega\,q, m/n) \approx \mathcal{H}om_{{}_q\mathcal{O}}(\Omega/q\,\Omega, m/n) \approx m/n \otimes_{{}_q\mathcal{O}} (\Omega/q\Omega)^*$$

by Prop. VIII.C.1 of G & R; here $(\Omega/q\,\Omega)^*$ is the dual sheaf to $\Omega/q\,\Omega$. Let Θ denote $(\Omega/q\Omega)^*$. Since the ambient sheaf of tangent vectors is dual to Ω, Θ is the sheaf of germs of sections of the restriction to A_2 of the ambient tangent bundle. Summarizing, we have an ${}_n\mathcal{O}$-module isomorphism

(6.3) $$_{n,m}\Theta \approx \Theta \otimes_{{}_q\mathcal{O}} m/n$$

As sheaves of ${}_q\mathcal{O}$-modules, the sheaves on the right are locally free.

In the case $\nu = 1$, we may repeat the above argument up to

$$\mathcal{H}om_{{}_n\mathcal{O}}({}_m\Omega/{}_m\Omega q, m/n) \approx \mathcal{H}om_{{}_q\mathcal{O}}({}_m\Omega/{}_m\Omega\, q, m/n)$$

and ${}_m\Omega/{}_m\Omega\,q \approx \Omega/\{$the $\mathcal{O}$-module generated by $q\,\Omega$ and $\Omega'\}$. However, $\Omega' \not\subset q\,\Omega$. Choose local coordinates with $A_1 = \{x=0\}$ and $A_2 = \{y=0\}$. Then $d(x^s y) - syx^{s-1}dx = x^s dy$ is in the $\mathcal{O}$-module generated by $q\,\Omega$ and Ω'. Given $\alpha \in \mathcal{H}om_{{}_q\mathcal{O}}({}_m\Omega/{}_m\Omega\, q, m/n)$, what is $\alpha(dy)$? If $\alpha(dy) \neq 0$, then $x^s\alpha(dy) \neq 0$ since m/n is a locally free ${}_y\mathcal{O}$-module. But $x^s\alpha(dy) = \alpha(x^s dy) = 0$ since α vanishes on the $\mathcal{O}$-module generated by $q\,\Omega$ and Ω'. Hence $\alpha(dy) = 0$. Thus α induces an ${}_q\mathcal{O}$-homomorphism in $\mathcal{H}om_{{}_q\mathcal{O}}(q\,\Omega, m/n)$. Summarizing, if $\nu = 1$, we have an ${}_n\mathcal{O}$-module isomorphism

$$(6.4) \qquad {}_{n,m}\Theta \approx \mathcal{H}om_{{}_q\mathcal{O}}({}_q\Omega, m/n) \approx {}_q\Theta \otimes {}_q\mathcal{O}^{m/n} .$$

Again, the right side involves only free ${}_q\mathcal{O}$-modules.

Let us now calculate $\mathcal{A}n(n,m)$. First we show that $\mathcal{A}n(n,m) = 1$ if $\nu \geq 2$. Use local coordinates as above. $\xi \in \mathcal{A}n(n,m)_o$ is locally induced by $(x,y) \to (u(x,y), v(x,y))$ in $\mathcal{A}ut(n{:}m)$ by Theorem 6.2. $\mathcal{A}n(n,m)$ may be thought of as those automorphisms of m/n which may be induced from $\mathcal{A}ut(n{:}m)$. ${}_m\mathcal{O} = \mathcal{O}/x^s y^\nu \mathcal{O}$ is mapped to itself identically. Hence Hence $u(x,y) = x + u_1$ with $u_1 \in x^s y^\nu \mathcal{O}$ and $v(x,y) = y + v_1 = y(1 + v_2)$ with $v_2 \in x^s y^{\nu - 1}\mathcal{O}$. In $m/n = x^s y^\nu \mathcal{O}/x^s y^{\nu+1}\mathcal{O}$, $x^s y^\nu f(x,y)$ has a representative of the form $x^s y^\nu \Sigma\, a_i x^i$. $x^{s+i}y^\nu$ is mapped to $(x+u_1)^{s+i}y^\nu(1+v_2)^\nu$. y divides both u_1 and v_2. x^s divides u_1 and v_2, so modulo $n = x^s y^{\nu+1}\mathcal{O}$, the induced map is the identity.

If $\nu = 1$, $s \neq 0$, $x^s y\; f(x,y) \equiv x^s y\; \Sigma\, a_i x^i$ has $x^{s+i}y$ mapped to $(x+u_1)^{s+i}y(1+v_2)$ where yx^s divides u_1 and x^s divides v_2. Hence

$$(x+u_1)^{s+i}y(1+v_2) \equiv x^{s+i}y(1+v_2) \bmod x^s y^{\nu+1} .$$

$v_2 = x^s h(x,y)$. $\tilde{h}(x) = h(x,0)$ determines the automorphism and different $\tilde{h}(x)$ determine different automorphisms. Hence $\mathcal{A}n(n,m)$ is in one-to-one correspondence with the functions $1 + x^s\tilde{h}(x)$. Under the composition of automorphisms, $y \to y^* = y(1+x^s\tilde{h}(x)) \to y^{**} = y^*(1+x^s\tilde{g}(x)) = y(1+x^s\tilde{h}(x))(1+x^s\tilde{g}(x))$. So the group action corresponds to a subsheaf of $\mathcal{O}^*$ which we shall call $\mathcal{O}_s^*$. Finally consider $\nu = 1$, $s = 0$,

$u(x,y) = x + yu_2$, $v(x,y) = y(1+v_2) = yv_3$ is an ambient isomorphism since it induces an isomorphism on $A(n)$, which is neatly embedded. Hence $v_3(0,0) \neq 0$ and $\mathfrak{An}(n,m)$ corresponds to all of $\mathcal{O}^*$. It is straightforward to verify that the isomorphism $\mathcal{O}_s^* \approx \mathfrak{An}(n,m)$ is independent of the choice of local coordinates.

Let $p' = p/pq$, i.e., the sheaf of germs of holomorphic functions on A_2 which vanish at $A_1 \cap A_2$. Let Z' be the subsheaf of the constant sheaf of integers over A_2 which has the zero stalk at $A_1 \cap A_2$.

$$0 \to Z' \to (p')^s \overset{\exp}{\to} \mathcal{O}_s^* \to 1$$

is then exact..

VIb. Cohomology in Sheaves of Non-abelian Groups

Let $\mathcal{F}$ be a sheaf of non-abelian groups over a paracompact Hausdorff topological space X. We wish to define cohomology with values in $\mathcal{F}$ for dimensions 0 and 1. $H^0(X,\mathcal{F}) = \Gamma(X,\mathcal{F})$ and $H^0(X,\mathcal{F})$ thus has a group structure. $H^1(X,\mathcal{F})$, defined below, will be a set with a distinguished element. Such a set is sometimes called a pointed set. We shall use $*$ to denote the distinguished element.

Let $\mathfrak{U} = \{U_i\}$ be a covering of X by open sets. $C^1(N(\mathfrak{U}),\mathcal{F})$, the set 1-cochains of $N(\mathfrak{U})$, the nerve of the cover $\mathfrak{U}$, with values in $\mathcal{F}$ is given by specifying for each $f \in C^1(N(\mathfrak{U}),\mathcal{F})$ a section $f_{ij} \in \Gamma(U_i \cap U_j,\mathcal{F})$ for every 1-simplex $U_i \cap U_j$. $C^1(N(\mathfrak{U}),\mathcal{F})$ has the obvious group structure. If $\mathfrak{V} < \mathfrak{U}$, i.e., $\mathfrak{V} = \{V_k\}$ is a refinement of $\mathfrak{U}$, then there is a natural restriction homomorphism $\rho: C^1(N(\mathfrak{U}),\mathcal{F}) \to C^1(N(\mathfrak{V}),\mathcal{F})$. The coverings of X form a directed set under the refinement relation. $C^1(X,\mathcal{F})$, the cochains of X with values in $\mathcal{F}$ is the direct limit of the directed system $\{C^1(N(\mathfrak{U}),\mathcal{F})\}$. $1: U_i \cap U_j \to 1$, i.e., the identity section, is a canonical distinguished cochain. The 0-cochains are defined similarly.

$Z^1(X,\mathfrak{F})$, the set of 1-cocycles, is that subset of $C^1(X,\mathfrak{F})$ given by those cochains [f] such that for some $f \in [f]$, $f_{ij}f_{jk}f_{ki} = 1$ on all $U_i \cap U_j \cap U_k \neq \emptyset$. We now define an equivalence relation on $Z^1(X,\mathfrak{F})$. Two cocycles [f] and [g] are equivalent if there is a 0-cochain [r] and some covering $\mathfrak{U}$ such that on $N(\mathfrak{U})$ there are representatives $f \in [f]$, $g \in [g]$ and $r \in [r]$ such that $f_{ij} = r_i^{-1} g_{ij} r_j$, where r_i and r_j are restricted to $U_i \cap U_j$. It is easy to check that this is an equivalence relation. The equivalence classes form $H^1(X,\mathfrak{F})$, the first cohomology set of X with values in $\mathfrak{F}$. The equivalence class $*$ containing 1 is called the coboundaries and is the distinguished element in $H^1(X,\mathfrak{F})$.

DEFINITION 6.3. *$\lambda: A \to B$ is a mapping of pointed sets if A and B are pointed sets, λ is a set mapping and $\lambda(*) = *$. A sequence of pointed set mappings*

$$A \xrightarrow{\lambda} B \xrightarrow{\tau} C$$

is exact at B if $\tau^{-1}(*) = \lambda(A)$.

THEOREM 6.5. *If $1 \to \mathfrak{R} \xrightarrow{\lambda} \mathfrak{S} \xrightarrow{\tau} \mathfrak{T} \to 1$ is an exact sequence of sheaves of non-abelian groups over a paracompact Hausdorff space X, then*

$$1 \to \Gamma(X,\mathfrak{R}) \xrightarrow{\lambda_0} \Gamma(X,\mathfrak{S}) \xrightarrow{\tau_0} \Gamma(X,\mathfrak{T}) \xrightarrow{\delta}$$

$$H^1(X,\mathfrak{R}) \xrightarrow{\lambda_1} H^1(X,\mathfrak{S}) \xrightarrow{\tau_1} H^1(X,\mathfrak{T})$$

is an exact sequence of pointed sets. δt is defined as follows. $t = \tau(s_i)$, $s_i \in \Gamma(U_i,\mathfrak{S})$ for some open cover $\mathfrak{U} = \{U_i\}$ of X. $(\delta t)_{ij} = \lambda^{-1}(s_i^{-1} s_j)$ determines an equivalence class $\delta t \in H^1(X,\mathfrak{R})$.

Proof: Exactness at $\Gamma(X,\mathfrak{R})$ is just that λ_0 is an injection.

$\tau_0\lambda_0 = 1$. If $\lambda_0(s) = 1$, s is in fact locally in $\mathfrak{R}$ and hence gives a section of $\mathfrak{R}$.

$\delta\tau_o(s)_{ij} = \lambda^{-1}(s^{-1}s) = 1$. If $\delta(t) = *$, after lifting t to $s_i \in \Gamma(U_i . \mathcal{S})$ locally, after suitable refinement, on $U_i \cap U_j$, $\lambda^{-1}(s_i^{-1}s_j) = r_i^{-1}r_j$, $r_i \in \Gamma(U_i, \mathcal{R})$. $s_j r_j^{-1} = s_i r_i^{-1}$ on $U_i \cap U_j$. Hence $\{s_i r_i^{-1}\}$ defines a section $b \in \Gamma(X, \mathcal{S})$ and $\tau(b) = t$.

$\lambda_1\delta(t) = \lambda(\lambda^{-1}(s_i^{-1}s_j) = s_i^{-1}s_j$, which is precisely the condition for $\lambda_1\delta(t)$ to be a coboundary. If $\lambda_1(r) = *$, for a suitable representative $\lambda(r_{ij}) = s_i^{-1}s_j$. $s_i\lambda(r_{ij}) = s_j$. $\tau(s_i\lambda(r_{ij})) = \tau(s_i)\tau\lambda(r_{ij}) = \tau(s_i) = \tau(s_j)$ on $U_i \cap U_j$. Hence the $\tau(s_i)$ define a section $t \in \Gamma(X, \mathcal{T})$ with $\delta t = r$.

$\tau_1\lambda_1(d)$ is the equivalence class of 1, i.e., $*$. If $\tau_1(s) = *$, for suitable representatives $\tau(s_{ij}) = t_i^{-1}t_j$ on $U_i \cap U_j$. After suitable refinement, $t_i = \tau(\sigma_i)$, $\sigma_i \in \Gamma(U_i, \mathcal{S})$. $\tau(s_{ij}) = \tau(\sigma_i)^{-1}\tau(\sigma_j)$. $\tau(\sigma_i s_{ij}\sigma_j^{-1}) = 1$. Hence $\sigma_i s_{ij}\sigma_j^{-1} = \lambda(r_{ij})$ for some $r_{ij} \in \Gamma(U_i \cap U_j, \mathcal{R})$. But $\sigma_i s_{ij}\sigma_j^{-1}$ determines the same equivalence class as s_{ij}, namely s. Hence $s = \lambda[(r_{ij})]$.∎

VIc. Formal Isomorphisms

Returning to our usual situation, let A be a 1-dimensional (reduced) subset of the 2-dimensional manifold M. Suppose that $\tilde{A}$ is a (reduced) 1-dimensional subset of the 2-dimensional manifold $\tilde{M}$ and $\tilde{n} \subset \tilde{m}$ are coherent ideal sheaves on $\tilde{M}$ with $\tilde{A}$ as their zero locus. Suppose $(\theta, \psi): A(m) \to \tilde{A}(\tilde{m})$, $\theta: A \to \tilde{A}$, $\psi: {}_{\tilde{m}}\mathcal{O} \to \theta_*({}_m\mathcal{O})$, is an isomorphism which can locally be extended to an isomorphism $A(n) \to \tilde{A}(\tilde{n})$, i.e., to each $x \in A$, there is a neighborhood U in A of x and an isomorphism $(\theta|_U, \phi_U): (U, {}_n\mathcal{O}) \to (\tilde{U}, {}_{\tilde{n}}\mathcal{O})$ such that $\theta^* \circ \phi_U(\tilde{m}/\tilde{n}) = m/n$ and $(\theta|_U, \phi_U)$ induces (θ, ψ) on $U(m)$. Theorems 6.1 and 6.2 insure that locally (θ, ϕ) can be extended to an ambient isomorphism. Thus ϕ_U will exist for similarly invariantly defined n and $\tilde{n}$, e.g., for the $m = p^s q^\nu$, $n = p^s q^{\nu+1}$ of VIa.

We shall now frequently abuse our notation by letting ψ stand for (θ, ψ).

THEOREM 6.6. *Let* $\psi: A(m) \to (\tilde{A}(\tilde{m})$ *be an isomorphism that can locally be extended to an isomorphism* $A(n) \to \tilde{A}(\tilde{n})$. ψ *determines a class* $[\psi] \in H^1(A, \mathcal{A}ut(n:m))$. $[\psi] = *$ *if and only if there is a global isomorphism* $\phi: A(n) \to \tilde{A}(\tilde{n})$ *such that* ϕ *induces* ψ.

Proof: Let $\mathfrak{U} = \{U_i\}$ be a covering of A such that ψ extends to ϕ_i on U_i. $\phi_i^{-1} \circ \phi_j$ then maps $(U_i \cap U_j)(n)$ isomorphically onto itself and induces the identity on $(U_i \cap U_j)(m)$. Thus $\phi_i^{-1} \circ \phi_j$ determines a class in $Z^1(N(\mathfrak{U}), \mathcal{A}ut(n:m))$. Let χ_i, defined on U_i (after suitable refinement) be a possibly different extension of ψ. $\chi_i^{-1} \circ \chi_j =$ $(\phi_i^{-1} \circ \chi_i)^{-1} \circ (\phi_i^{-1} \circ \phi_j) \circ (\phi_j^{-1} \circ \chi_j)$ so $(\chi_i^{-1} \circ \chi_j)$ and $(\psi_i^{-1} \circ \psi_j)$ determine the same cohomology class $[\psi]$.

If there is a global ϕ, let ϕ_i be ϕ restricted to U_i. Then $[\psi] = [\phi^{-1} \circ \phi] = [1] = *$.

If $[\psi] = *$, after suitable refinement, $\phi_i^{-1} \circ \phi_j = \psi_i^{-1} \circ \psi_j$ for a suitable cochain $(\psi_i) \in C^1(N(\mathfrak{U}), \mathcal{A}ut(n:m))$. Then $\phi_j \circ \psi_j^{-1} = \phi_i \circ \psi_i^{-1}$ on $U_i \cap U_j$. Hence the $\phi_i \circ \psi_i^{-1}$ patch together to give an isomorphism $\phi: A(n) \to \tilde{A}(\tilde{n})$. Since ψ_i induces the identity on $A(m)$, ϕ induces ψ. ∎

THEOREM 6.7. *If* $H^1(A, \mathcal{A}n(n,m)) = H^1(S,_{n,m}\Theta) = *$, *then* $H^1(A, \mathcal{A}ut(n:m)) = *$.

Proof: This follows immediately from (6.2), Proposition 6.4 and Theorem 6.5. ∎

The idea now is to consider two exceptional sets A and $\tilde{A}$ embedded in M and $\tilde{M}$ respectively. Suppose that A and $\tilde{A}$ are isomorphic as reduced analytic spaces. To determine if A and $\tilde{A}$ have biholomorphic neighborhoods, we may try to extend the original isomorphism $\phi_o\colon A \to \tilde{A}$ to isomorphisms between non-reduced spaces with A and $\tilde{A}$ as the underlying topological spaces. Using Theorem 6.6, we shall show that once ϕ_o has been extended to certain non-reduced spaces, it may be extended indefinitely. In the last part of this section, we shall show that in such cases, A and $\tilde{A}$ do have biholomorphic neighborhoods.

Let $A = \cup A_i$, $\tilde{A} = \cup \tilde{A}_i$ be the decomposition of A and $\tilde{A}$ into irreducible components. As usual, we assume that the A_i are non-singular, intersect transversely, and no three meet at a point. Let p_i be the ideal sheaf of A_i. We shall only consider cases where $m = \Pi\, p_i^{s_i}$, $n = \Pi\, p_i^{s_i'}$ with $s_i = s_i'$ for all i except one, say i_o, and $s_{i_o}' = s_{i_o} + 1$. If $s_{i_o} \geq 2$, $\mathcal{A}n(n,m) = 1$ so to apply Theorem 6.7 and Theorem 6.6, it suffices, (6.3), that

$$H^1(A,{}_{n,m}\Theta) \approx H^1(A_{i_o}, \Theta \otimes m/n) = 0\ .$$

If $\mathcal{T}$ is the sheaf of germs of sections of the tangent bundle of the reduced space A_{i_o} and $\mathcal{N}$ is the sheaf of germs of sections of the normal bundle of the embedding,

$$0 \to \mathcal{T} \to \Theta \to \mathcal{N} \to 0$$

is exact. These sheaves are locally free sheaves over the reduced space, so

$$0 \to \mathcal{T} \otimes m/n \to \Theta \otimes m/n \to \mathcal{N} \otimes m/n \to 0$$

is also exact.

$$H^1(A_{i_o}, \Theta \otimes m/n) = 0$$

if

$$H^1(A_{i_o}, \mathcal{T} \otimes m/n) = H^1(A_{i_o}, \mathcal{N} \otimes m/n) = 0\ .$$

$\mathcal{T} \otimes m/n$ and $\mathfrak{N} \otimes m/n$ are of rank 1. If $\mathcal{F}$ is a locally free sheaf of rank 1 over A_{i_o}, let $c(\mathcal{F})$ be the Chern class of the corresponding line bundle. Let g be the genus of A_{i_o} and κ be the canonical bundle, i.e., the cotangent bundle; Ω^1 is the sheaf of germs of sections of κ. $c(\kappa) = 2g-2$. $c(\mathcal{T}) = 2-2g$. For simplicity of notation, let $i_o = 1$. $c(\mathfrak{N}) = A_1 \cdot A_1$ (the self-intersection number, see Theorem 2.3).

$$c(p_1^{s_1}/p_1^{s_1+1}) = c(\underset{s_1}{\otimes} N^*) = -s_1\, c(N) = -A_1 \cdot (s_1A_1)\ .$$

m/n corresponds to those sections of $\underset{s_1}{\otimes} N^*$ which vanish to order s_j on $A_j \cap A_1$, $j \neq 1$. Hence $c\,(m/n) = -A_1 \cdot (\Sigma\, s_iA_i)$. Hence $c(\mathcal{T} \otimes m/n) = 2-2g - A_i \cdot\ (\Sigma\, s_iA_i)$ and $c(\mathfrak{N} \otimes m/n) = A_1 \cdot A_1 - A_1 \cdot (\Sigma\, s_iA_i)$. Since by Serre duality [Gu, p. 95], for any locally free sheaf $\mathcal{F}$ of rank 1, $H^1(A_1, \mathcal{F}) \approx [\Gamma(A_1, \Omega^1 \otimes \mathcal{F}^*)]^*$, $H^1(A_1, \mathcal{F}) = 0$ if $c(\Omega^1 \otimes \mathcal{F}^*) = 2g-2-c(F) < 0$. Thus

THEOREM 6.8. *An isomorphism* $\psi: A(m) \to \tilde{A}(\tilde{m})$ *may be extended to an isomorphism* $\phi: A(n) \to \tilde{A}(\tilde{n})$ *if* $2(2g-2) + A_{i_o} \cdot\ (\Sigma\, s_iA_i) < 0$ *and* $2g-2-A_{i_o} \cdot A_{i_o} + A_{i_o} \cdot (\Sigma\, s_iA_i) < 0$.

We shall now use the negative definiteness of $(A_i\ .\ A_j)$ (Theorem 4.4) to find an $m = \Pi\, p_i^{s_i}$ such that for suitably chosen $m^{(0)} = m, m^{(1)}, m^{(2)}, \ldots$ with $\min_i s_i^{(\nu)} \to \infty$ as $\nu \to \infty$, $m^{(\nu+1)}$ and $m^{(\nu)}$ will satisfy the hypotheses of Theorem 6.8. Thus if $A(m) \approx \tilde{A}(\tilde{m})$, we will automatically an isomorphism $A(n) \approx \tilde{A}(\tilde{n})$ where n may be any power of the ideal sheaf of A. By Lemma 4.10, there exist natural numbers $(r_1, \ldots, r_n)$ such that $A_i \cdot \sum_j r_jA_j = \sum_j r_jA_i \cdot A_j \leq -1$ for all i. These r_i will now be fixed throughout this section. Let $Z = \Sigma r_iA_i$, so that Z is a cycle such that $A_j \cdot Z \leq -1$ for all j. Let $\mathcal{J} = \Pi_i^{r_i}$ be the corresponding ideal sheaf. Choose cycles $Z_1, Z_2, \ldots, Z_{r-1}, Z_r = Z$, $r = \Sigma r_i$, such that

$Z_1 = A_{i_o}$, any A_{i_o}, and $Z_{k+1} - Z_k = A_{i_k}$ for some irreducible component A_{i_k}. The Z_k corresponds to ideal sheaves $\mathcal{J}_k$ such that $\mathcal{J}_{k+1} = p_{i_k}\mathcal{J}_k$. $c(\mathcal{J}_k/\mathcal{J}_{k+1}) = - A_{i_k} \cdot Z_k$ by the computation preceding Theorem 6.8. Let τ be the maximum of the $A_{ij} \cdot Z_j$. Let λ be the maximum of $\{0, \{2(2g_j-2)\}, \{2g_j-2-(A_j \cdot A_j)\}\}$, where g_j is the genus of A_j.

THEOREM 6.9. *Any isomorphism* $\psi: A(\mathcal{J}^\nu) \to \tilde{A}(\mathcal{J}^\nu)$ *with* $\nu \geq \lambda + \tau + 1$ *may be extended to an isomorphism* $\phi: A(\mathcal{J}^{\nu+1}) \to \tilde{A}(\tilde{\mathcal{J}}^{\nu+1})$.

Proof: We apply Theorem 6.8 successively to $m^{(0)} = \mathcal{J}^\nu$, $m^{(1)} = \mathcal{J}^\nu \mathcal{J}_1$, $m^{(2)} = \mathcal{J}^\nu \mathcal{J}_2, \ldots, m^{(r)} = \mathcal{J}^\nu \mathcal{J} = \mathcal{J}^{\nu+1}$. $A_{i_k} \cdot (\Sigma s_i A_i) = A_{i_k} \cdot (\nu Z + Z_k) \leq \nu + A_{i_k} \cdot Z_k \leq -\nu + \tau \leq -\lambda - \tau - 1 + \tau = -\lambda - 1$. But $\lambda \geq 2(2g_{i_k} - 2)$ and $\lambda \geq 2g_{i_k} - 2 - A_{i_k} \cdot A_{i_k}$. Hence ψ extends to $\phi^{(1)}: A(\mathcal{J}^\nu \mathcal{J}_1) \to \tilde{A}(\tilde{\mathcal{J}}^\nu \tilde{\mathcal{J}}_1)$, $\phi^{(2)}, \ldots$ and up to $\phi^{(r)}: A(\mathcal{J}^{\nu+1}) \to \tilde{A}(\tilde{\mathcal{J}}^{\nu+1})$. ∎

We now introduce some unneeded machinery in order to try to clarify the ideas. The ideal sheaves $\mathcal{P}_s = \Pi\, p_i^{s_i}$, with s a multi-index, form a directed set under $\mathcal{P}_s < \mathcal{P}_{s'}$ if $s_i \leq s_i'$, all i. If $\mathcal{P}_s < \mathcal{P}_{s'}$, there is a canonical surjective mapping of structure sheaves $\mathcal{O}/\mathcal{P}_{s'} \to \mathcal{O}/\mathcal{P}_s$. Let ${}_\infty\mathcal{O}$ be the inverse limit of the system $\mathcal{O}/\mathcal{P}_s$. ${}_\infty\mathcal{O}$ has the natural structure of a sheaf over A. Near a regular point, for example, $a = (0,0)$, $A = \{y=0\}$, a cofinal set may be given by ${}_y\mathcal{O}$, ${}_{y^2}\mathcal{O}, \ldots$. Let ${}_1\mathcal{O}_a$ represent convergent power series in x. ${}_{y^2}\mathcal{O}_a \approx {}_1\mathcal{O}_a \oplus {}_1\mathcal{O}_a y$. ${}_{y^3}\mathcal{O}_a \approx {}_1\mathcal{O}_a \oplus {}_1\mathcal{O}_a y \oplus {}_1\mathcal{O}_a y^2$. ... Thus ${}_\infty\mathcal{O}_a$ may be thought of as formal power series in y with coefficients which are convergent power series in x. Thus ${}_\infty\mathcal{O}_a$ differs from $\mathcal{O}_a$, the stalk of the ambient structure sheaf, only in that its power series do not have to converge.

DEFINITION 6.4. *The pair* $A(\infty) = (A, {}_\infty\mathcal{O})$ *is called a formal neighborhood of* A. *Two reduced analytic subspaces* A *and* $\tilde{A}$ *of* M *and* $\tilde{M}$ *respectively are said to have formally equivalent neighborhoods if there is a* $\mathbb{C}$*-algebra isomorphism* $(\theta, \phi): A(\infty) \to \tilde{A}(\infty)$ *as in Definition 6.1.*

Since elements of ${}_\infty\mathcal{O}$ may be evaluated as functions, the ideal sheaves of the A_i are well defined in ${}_\infty\mathcal{O}$. Hence ϕ necessarily induces an isomorphism $\phi_s: A(\mathcal{P}_s) \to \tilde{A}(\tilde{\mathcal{P}}_s)$ for all s.

We may reformulate Theorem 6.9 as

THEOREM 6.10. *If* $\psi: A(\mathcal{J}^\nu) \to \tilde{A}(\tilde{\mathcal{J}}^\nu)$ *is an isomorphism,* $\nu \geq \lambda + \tau + 1$, *then* A *and* $\tilde{A}$ *have formally equivalent neighborhoods.*

We shall need some more results for later.

LEMMA 6.11. *Let* A *and* $\tilde{A}$ *be 1-dimensional reduced analytic subvarieties of the 2-dimensional manifolds* M *and* $\tilde{M}$. *Let the irreducible components of* A *and* $\tilde{A}$ *be non-singular, intersect transversely and no three meet at a point. Let* $\mathcal{J}$ *and* $\tilde{\mathcal{J}}$ *be the ideal sheaves of* A *and* $\tilde{A}$ *respectively. Let* $\pi: M' \to M$ *and* $\tilde{\pi}: \tilde{M}' \to M$ *be quadratic transformations at* p *and* $\tilde{p}$ *and let* $\mathcal{J}'$ *and* $\tilde{\mathcal{J}}'$ *be the ideal sheaves of* $A' = \pi^{-1}(A)$ *and* $\tilde{A}' = \tilde{\pi}^{-1}(\tilde{A})$ *respectively. If* $(\theta, \phi): A(\mathcal{J}^\nu) \to \tilde{A}(\tilde{\mathcal{J}}^\nu)$ *is an isomorphism with* $\theta(p) = \tilde{p}$, *then* (θ, ϕ) *canonically induces an isomorphism*
$(\theta', \phi'): A'((\mathcal{J}')^{\nu-1}) \to \tilde{A}((\tilde{\mathcal{J}}')^{\nu-1})$.

Proof: We may assume that $\nu \geq 2$ for if $\nu = 1$, $\mathcal{J}^{\nu-1} = \mathcal{O}$. (θ', ϕ') is induced in the usual manner off p. Near p, $A(\mathcal{J}^\nu)$ is neatly embedded so we may choose local coordinates (x,y) near p = (0,0) and $(\tilde{x}, \tilde{y})$ near $\tilde{p} = (0,0)$ such that (θ, ϕ) is induced by $\tilde{x} = x$, $\tilde{y} = x$ (Theorem 6.2). This ambient isomorphism certainly induces a (θ', ϕ') in a neighborhood of $\pi^{-1}(p)$, but we must verify that any other ambient map Φ which induces (θ, ϕ) induces the same (θ', ϕ'). Suppose $A = \{x = 0\}$. (The case where $A = \{xy = 0\}$ is essentially the same calculation.) Φ is given by $\tilde{x} = x + u(x,y)$, $\tilde{y} = y + v(x,y)$ with u and v in the ideal $x^\nu\mathcal{O}$.

π is given by $(x,y) = (\xi\zeta, \zeta) = (\zeta', \xi'\zeta')$ and $\tilde{\pi}$ by $(\tilde{x},\tilde{y}) = (\tilde{\xi}\tilde{\zeta}, \tilde{\zeta}) = (\tilde{\zeta}', \tilde{\xi}'\tilde{\zeta}')$. We must show that $\tilde{\zeta} = \zeta + r(\xi,\zeta)$, $\tilde{\xi} = \xi + s(\xi,\zeta)$ with r and s in $\xi^{\nu-1}\zeta^{\nu-1}\mathcal{O}$. $\tilde{\zeta} = \zeta + r(\xi,\zeta)$ and $\tilde{\xi} = \xi + s(\xi,\zeta)$ will then induce the same map on $A((\mathcal{J}')^{\nu-1})$ as does $\tilde{\zeta} = \zeta$ and $\tilde{\xi} = \xi$.

$\tilde{\zeta} = \tilde{y} = y + v(x,y) = \zeta + v(\xi\zeta,\zeta)$. $v(\xi\zeta,\zeta) \in \xi^{\nu}\zeta^{\nu}\mathcal{O}$ since $v(x,y) \in x^{\nu}\mathcal{O}$.

$$\tilde{\xi} = \tilde{x}/\tilde{y} = x + u(x,y)/y + v(x,y) = [\xi\zeta + u(\xi\zeta,\zeta)]/[\zeta + v(\xi\zeta,\zeta)]$$

$$= [\xi + u(\xi\zeta,\zeta)/\zeta] \;/\; [1 + v(\xi\zeta,\zeta)/\zeta]$$

$$= \xi + \frac{1}{\zeta}\, u(\xi\zeta,\zeta))\;(1 - \frac{1}{\zeta}\, v(\xi\zeta,\zeta) + [\frac{1}{\zeta}\, v(\xi\zeta,\zeta)]^2 - \dots)\,.$$

Since $u(x,y) = x^{\nu}u_1(x,y)$, $\frac{1}{\zeta}u(\xi\zeta,\zeta) = \xi^{\nu}\zeta^{\nu-1}u_1(\xi\zeta,\zeta)$. Similarly, $\frac{1}{\zeta}v(\xi\zeta,\zeta) \in \xi^{\nu}\zeta^{\nu-1}\mathcal{O}$. This completes this calculation and the remaining calculations are similar.▮

THEOREM 6.12. *Let* A *be a compact* 1-*dimensional reduced subvariety of* M, *a* 2-*dimensional manifold. Let the irreducible components of* A *be non-singular, intersect transversely, and no three meet at a point. Let* $\tilde{A}$ *be a subvariety of* $\tilde{M}$, *a* 2-*dimensional manifold. Let* $\mathcal{J}$ *and* $\tilde{\mathcal{J}}$ *be the ideal sheaves of* A *and* $\tilde{A}$ *respectively. If* A *is exceptional in* M *and there is an isomorphism* $\phi: A(\mathcal{J}^2) \to \tilde{A}(\tilde{\mathcal{J}}^2)$, *then* $\tilde{A}$ *is exceptional in* $\tilde{M}$. A *and* $\tilde{A}$ *also have the same intersection matrix.*

Proof: ϕ induces an isomorphism between the reduced spaces A and $\tilde{A}$ so $\tilde{A}$ is also 1-dimensional with non-singular irreducible components. $\tilde{A}$ must have its irreducible components crossing transversely, for let $\tilde{p}$ be a singular point of $\tilde{A}$. Perform quadratic transformations at p and $\tilde{p}$. By Lemma 6.11, ϕ induces an isomorphism on the reduced spaces A' and $\tilde{A}'$. Since the irreducible components of A cross transversely, they meet $\pi^{-1}(p)$ at distinct points. Hence the irreducible components of $\tilde{A}$ meet $\tilde{\pi}^{-1}(\tilde{p})$ at distinct points and hence cross transversely in $\tilde{M}$. Thus $A_i \cdot A_j = \tilde{A}_i \cdot \tilde{A}_j$, $i \neq j$. $A_i \cdot A_i$ is the Chern class of the normal bundle of the embedding and is thus determined by the structure of p_i/p_i^2 as a sheaf of ${}_{p_i}\mathcal{O}$-modules. But $\mathcal{O}/\mathcal{J}^2$ has $\mathcal{O}/p_i$ as a canonical quotient sheaf and p_i/p_i^2 as a canonical subsheaf of a canonical quotient sheaf. Thus $\mathcal{O}/\mathcal{J}^2$ does determine the structure of p_i/p_i^2 as a sheaf of ${}_{p_i}\mathcal{O}$-modules. Hence ϕ establishes that $A_i \cdot A_i = \tilde{A}_i \cdot \tilde{A}_i$. Then by Theorem 4.9, $\tilde{A}$ is exceptional since A is exceptional. ▮

VId. Formal Equivalence Implies Actual Equivalence

THEOREM 6.13. *Let $\phi : A(\infty) \to \tilde{A}(\infty)$ be an isomorphism showing that A and $\tilde{A}$, 1-dimensional reduced analytic subsets of the complex manifolds M and $\tilde{M}$, have formally equivalent neighborhoods. Suppose that A is exceptional in M and its irreducible components are non-singular, cross transversely, and no three meet at a point. Then there are neighborhoods U and $\tilde{U}$ of A and $\tilde{A}$ respectively and a biholomorphic map $\psi : U \to \tilde{U}$ such that ψ and ϕ induce the same map between the reduced spaces A and $\tilde{A}$.*

We do not claim that ψ induces ϕ. The proof of Theorem 6.13 will take quite some time.

Let $\Phi: M \to Y$ exhibit A as an exceptional set. Theorem 6.13 depends only on small neighborhoods of A so we may assume that Y is a subvariety of a polydisc Δ in C^n with $\Phi(A) = 0$ and Y neatly embedded in Δ at the origin. Let $\mathcal{T}$ be the sheaf of germs of sections of T, the tangent bundle of C^n. Let N , defined on Y-0, be the normal bundle of the embedding of Y-0 in Δ-0.

PROPOSITION 6.14. *Choosing, if necessary, a smaller polydisc* Δ, *there are a holomorphic function* g *on* Y, *such that* $g(0) = 0$, $g \not\equiv 0$, g *generates* $\mathcal{I}d(V(g))_z$ *for* $z \neq 0$, *and sections* $s_1, \ldots, s_{n-2} \in \Gamma(\Delta, \mathcal{T})$ *such that for* $z \in Y - V(g)$, *the images in* N_z *of* $s_1(z), \ldots, s_{n-2}(z)$ *form a basis of* N_z.

Proof: If $n = 2$ the proposition is trivial, so we may assume that 0 is a singular point of Y.

Choose coordinates and a small polydisc neighborhood of 0 such that $\pi: Y \to \Delta' \subset \mathbb{C}^2$ is an admissible representation. Let $B \subset \Delta'$ be the branch locus of π. B is then 1-dimensional so upon further restriction, we may assume that the origin is the only singularity of B. Let $p_1, \ldots, p_t \in B$ be points in each of the irreducible components of B near the origin, with no $p_i = 0$. Let $\{q_1, \ldots, q_s\} = \pi^{-1}(p_1, \ldots, p_t)$. If $\{f_{ik}\}$ generate $(\mathcal{I}d\, Y)_{q_i}$, $1 \leq k \leq n-2$, then $\{df_{ik}\}$ generate an (n-2)–dimensional subspace of the cotangent vectors to $\mathbb{C}^n$ at q_i. Let

$$\{s_1, \ldots, s_e, \ldots, s_{n-2}\} = \{ \sum_{j=1}^{n} t_{ej} \frac{\partial}{\partial z_j} \},$$

where the t_{ej} are complex constants, be chosen so that the matrix $<(df_{ik})_{q_i}, t_e>$ has rank n–2 for each i. Let $V = \{z \in Y \mid$ if $f_1, \ldots, f_r$ generate $(\mathcal{I}d\, Y)_z$, then the $r \times (n-2)$ matrix $<df_k, s_e>$ has rank less than $n-2\}$. V is then a proper subvariety of Y. Since $p_i \notin \pi(V)$, $\pi(V)$ is a proper subvariety of Δ' which contains no irreducible components of B. Then $\pi(V) \cap B$ has dimension 0 and hence is a discrete set. $0 \in V$ since 0 is a singular point. Thus after additional restriction, we may assume that $\pi(V) \cap B = 0$. If $\pi(V)$ is of codimension 1, by Lemma VIII.B.12 of G & R, there is a holomorphic function h on Δ' which generates $\mathcal{I}d(\pi)V))$ at each point. Since π is locally biholomorphic off $\pi^{-1}(B)$ and $\pi(V) \cap B = 0$, $g = \pi^*(h)$ generates $\mathcal{I}d(V(g))$ at each $z \in Y - 0$.

If $\pi(V)$ is of codimension 2, i.e., $\pi(V) = 0$, let h be any function on Δ' such that $B \cap V(h) = 0$. Then repeat the last argument.∎

Now g, chosen as in Proposition 6.14, induces $\Phi^*(g)$, also denoted g, a holomorphic function on M in a neighborhood of A. $g(A) = 0$. Perform quadratic transformations in M at points of A until near A, V(g) has non-singular irreducible components which intersect transversely with no three meeting at a point. By Lemma 6.11 A' and $\tilde{A}'$, the results of the quadratic transformations on M and on $\tilde{M}$ at the corresponding points, have formally equivalent neighborhoods if A and $\tilde{A}$ have formally equivalent neighborhoods. A' remains exceptional in M' so all the hypotheses of Theorem 6.13 are fulfilled. It suffices to show that A' and $\tilde{A}'$ have biholomorphic neighborhoods for then we may blow down the results of the quadratic transformations and establish the desired isomorphism. Hence, without loss of generality, we may omit the primes and assume that V(g) has the usual nice properties near A.

Let $D = \cup D_k$ be the union of those irreducible components D_k of V(g) such that $D_k \not\subset A$.

We need a result like Theorem 6.9 and Lemma 5.6.

Let F be a line bundle over U, a holomorphically convex neighborhood of A. Let $\mathcal{F}$ be its sheaf of germs of sections and $c_i(\mathcal{F}) = c(F \mid A_i)$. Recall the definitions of the $\mathcal{I}_k$ given before Theorem 6.9. Let $\Phi : U \to Y$ represent A as exceptional, with Y a Stein space.

LEMMA 6.15. *Let $\mathcal{F}$ be a locally free sheaf of rank 1 over U, a holomorphically convex neighborhood of A. Let $\sigma = \max\{2g_j - 2 + A_{i_j} \cdot Z_j\}$ If $c_j(\mathcal{F}) > \sigma$ for all j, if $\mathcal{F}$ is a subsheaf of a free sheaf, and if $\Phi_*(\mathcal{F})$ is coherent, then $H^1(U,\mathcal{F}) = 0$.*

Proof: Consider successively

$$0 \to \mathcal{I}_{k+1}\mathcal{F} \to \mathcal{I}_k\mathcal{F} \to \mathcal{I}_k\mathcal{F}/\mathcal{I}_{k+1}\mathcal{F} \to 0 \tag{6.5}$$

$\mathcal{I}_k\mathcal{F}/\mathcal{I}_{k+1}\mathcal{F}$ is a locally free sheaf of rank 1 over the reduced space A_{i_k}. If B is the bundle corresponding to $\mathcal{I}_k/\mathcal{I}_{k+1}$, $\mathcal{I}_k\mathcal{F}/\mathcal{I}_{k+1}\mathcal{F}$ is the sheaf of germs of sections of $B \otimes (F \mid A_{i_k})$. Hence

$c(\mathcal{I}_k\mathcal{F}/\mathcal{I}_{k+1}\mathcal{F}) = c(B) + c(F \mid A_{i_k}) = - A_{i_k} \cdot Z_k + c_{i_k}(F) > 2g_{i_k} - 2.$

By Serre duality, $H^1(\mathcal{I}_k\mathcal{F}/\mathcal{I}_{k+1}\mathcal{F})$ is dual to the space of sections of a bundle of negative Chern class. Hence $H^1(\mathcal{I}_k\mathcal{F}/\mathcal{I}_{k+1}\mathcal{F}) = 0$. From (6.5), the map $H^1(U,\mathcal{I}_{k+1}\mathcal{F}) \to H^1(U,\mathcal{I}_k\mathcal{F})$ is surjective for all k. By composition, $H^1(U,\mathcal{I}\mathcal{F}) \to H^1(U,F)$ is surjective.

$$c_j(\mathcal{I}\mathcal{F}) = c_j(\mathcal{I}) + c_j(\mathcal{F}) = -A_j \cdot Z + c_j(\mathcal{F}) > c_j(\mathcal{F}) > \sigma.$$

$\Phi_*(\mathcal{I}\mathcal{F})$ is coherent by Lemma 5.2. Hence $\mathcal{I}\mathcal{F}$ satisfies the hypotheses of this lemma. Repeating the argument above, we see that for all ν, the map $\rho: H^1(U, \mathcal{I}^\nu\mathcal{F}) \to H^1(U, \mathcal{F})$ is surjective. By Theorem 5.4, ρ is the zero map for sufficiently large ν. Hence $H^1(U,\mathcal{F}) = 0$. ∎

By Lemma 6.15, for $\mu \geq \sigma + 1$, $H^1(U, \mathcal{I}^\mu) = 0$ and the restriction map $\Gamma(U, \mathcal{O}) \to \Gamma(U, \mathcal{O}/\mathcal{I}^\mu)$ is surjective. Since $\tilde{A}$ is exceptional in $\tilde{M}$ and has the same intersection matrix by Theorem 6.12, $\Gamma(\tilde{U}, \mathcal{O}) \to \Gamma(\tilde{U}, \mathcal{O}/\tilde{\mathcal{I}}^\mu)$ is surjective for any sufficiently small holomorphically convex neighborhood $\tilde{U}$ of $\tilde{A}$ in $\tilde{M}$. There is an isomorphism $\phi: \mathcal{O}/\tilde{\mathcal{I}}^\mu \approx \theta_*(\mathcal{O}/\mathcal{I}^\mu)$. Thus for any $\mu \geq \sigma + 1$ we may find a $\tilde{g}$ holomorphic on $\tilde{U}$ such that the images of g and $\tilde{g}$ in $\mathcal{O}/\mathcal{I}^\mu$ and $\mathcal{O}/\tilde{\mathcal{I}}^\mu$ correspond under the isomorphism ϕ. Let χ be the maximum of the orders to which g vanishes on $\{A_i, D_k\}$, the irreducible components of V(g). Let $\mu \geq \chi + 1$. $\tilde{g}(\tilde{A}) = 0$ since g vanishes on A. Let $\{\tilde{D}_k\}$ be the irreducible components of $V(\tilde{g})$ near $\tilde{A}$ such that $\tilde{D}_k \not\subset \tilde{A}_k$. The $\tilde{D}_k$, in fact, correspond naturally to the D_k, are non-singular, and intersect $\tilde{A}$ transversely. Also $\tilde{g}$ vanishes on the $\tilde{D}_k$ and $\tilde{A}_i$ to the same order as g vanishes on the corresponding D_k and A_i. We see this as follows. Let $g + \mathcal{I}^\mu$ denote the image of g in $\mathcal{O}/\mathcal{I}^\mu$. g vanishes to order ν_i on A_i if and only if $g+\mathcal{I}^{\mu} \in (p_i^{\nu_i}+\mathcal{I}^{\mu})-(p_i^{\nu_i+1}+\mathcal{I}^{\mu})$. Since g and $\tilde{g}$ have corresponding images in $\mathcal{O}/\mathcal{S}^{\mu}$, g and $\tilde{g}$ vanish to the same order on corresponding A_i. Let $\mathcal{I}(a)$ be the ideal sheaf of a point $a \in A$. Then $a \in A_i \cap D$ if and only if $g+\mathcal{I}^{\mu} \in p_i^{\nu_i}\mathcal{I}(a)+\mathcal{I}^{\mu}$. The D_k's meet A at distinct points so the points of $A \cap D$ correspond to the D_k. The corresponding points in $\tilde{A}$ must also be in $\tilde{A} \cap \tilde{D}$ but we still must show that the $\tilde{D}_k$ are non-singular, cross $\tilde{A}$ transversely and no two $\tilde{D}_k$ meet $\tilde{A}$ at the same point. Choose local coordinates (x,y) and $(\tilde{x},\tilde{y})$ near corresponding points in $A \cap D$ and $\tilde{A} \cap \tilde{D}$ such that $\theta^*\phi: \mathcal{O}/\tilde{\mathcal{I}}^{\mu} \to \mathcal{O}/\mathcal{I}^{\mu}$ is induced by $\tilde{x}=x$, $\tilde{y}=y$ and such that $A=\{y=0\}$ and $D=\{x=0\}$. g vanishes to first order on the D_k by Proposition 6.14. Thus $g(x,y)=y^{\nu}x\,u(x,y)$ with $\nu \le \chi \le \mu-1$ and $u(0,0) \neq 0$. $\tilde{g}(\tilde{x},\tilde{y})=\tilde{y}^{\nu}x\,u(\tilde{x},\tilde{y})+\tilde{y}^{\mu}h(\tilde{x},\tilde{y})$ for some holomorphic function h. Near $\tilde{a}$, $\tilde{D}$ is the locus of $\tilde{x}\,u(\tilde{x},\tilde{y})+\tilde{y}^{\mu-\nu}h(\tilde{x},\tilde{y})$ and $\tilde{A}=\{\tilde{y}=0\}$. But $\frac{\partial}{\partial\tilde{x}}(\tilde{x}\,u(\tilde{x},\tilde{y})+\tilde{y}^{\mu-\nu}h(\tilde{x},\tilde{y}))$ $=\tilde{u}(\tilde{x},\tilde{y})+\tilde{x}\frac{\partial u}{\partial\tilde{x}}+\tilde{y}^{\mu-\nu}\frac{\partial h}{\partial\tilde{x}}$ equals $u(0,0) \neq 0$ at $\tilde{x}=0$, $\tilde{y}=0$. Thus $\tilde{D}$ is a submanifold near $\tilde{a}$ which meets $\tilde{A}$ transversely and $\tilde{g}$ vanishes to first order on D (for otherwise $\frac{\partial\tilde{g}}{\partial\tilde{x}}$ would vanish on D). $\pi(\tilde{x},\tilde{y})=\tilde{y}$ induces a biholomorphic map on $\tilde{D}$ for sufficiently small $\tilde{x}$ and $\tilde{y}$. Since on $\tilde{D}$, $\tilde{x}=-\tilde{y}^{\mu-\nu}h/u$, letting $\pi^{-1}(\tilde{y})=(r(\tilde{y}),\tilde{y})$ gives $r(\tilde{y})$ a $(\mu-\nu)^{th}$ order 0 at $\tilde{y}=0$. $\mu-\nu \ge \mu-\chi$.

Let q be the ideal sheaf of D and q the ideal sheaf of D and $\tilde{q}$ the ideal sheaf of $\tilde{D}$. Given γ, we wish to extend the isomorphism $\phi: A(\mathcal{I}^{\mu-\chi}) \to \tilde{A}(\mathcal{I}^{\mu-\chi})$ to $\phi^+: (A \cup D)(\mathcal{I}^{\mu-\chi}q^{\gamma}) \to (\tilde{A} \cup \tilde{D})(\tilde{\mathcal{I}}^{\mu-\chi}\tilde{q}^{\gamma})$. Near $(x,y)=0$, we induce ϕ^+ by $y=\tilde{y}$, $x=\tilde{x}-r(\tilde{y})$. We must check that ϕ^+ induces ϕ. But $r(\tilde{y})=\tilde{y}^{\mu-\chi}s(\tilde{y})$ so $r(\tilde{y}) \in \mathcal{I}^{\mu-\chi}$ near (0,0) and ϕ^+ does induce ϕ (which was induced by $y=\tilde{y}$, $x=\tilde{x}$). We shall always take $\gamma \ge 2$.

Let ω be the number of points in $A \cap D$. Recall the definitions of λ and τ from Theorem 6.9.

PROPOSITION 6.16. *Any isomorphism* $\psi: (A \cup D)(\mathcal{I}^{\nu} q^{\gamma}) \to (\tilde{A} \cup \tilde{D})(\tilde{\mathcal{I}}^{\nu} \tilde{q}^{\gamma})$ *with* $\nu \geq \lambda + \tau + \gamma\omega + 1$ *may be extended to an isomorphism* $\phi: (A \cup D)(\mathcal{I}^{\nu+\omega} q^{\gamma+1}) \to (\tilde{A} \cup \tilde{D})(\tilde{\mathcal{I}}^{\nu+\omega} \tilde{q}^{\gamma+1})$.

Proof: We first apply Theorem 6.8 successively to $m^{(0)} = \mathcal{I}^{\nu} q^{\gamma}$, $m^{(1)} = \mathcal{I}^{\nu} \mathcal{I}_1 q^{\gamma}$, $m^{(2)} = \mathcal{I}^{\nu} \mathcal{I}_2 q^{\gamma}, \ldots, m^{(r)} = \mathcal{I}^{\nu} \mathcal{I} q^{\gamma} = \mathcal{I}^{\nu+1} q^{\gamma}$. $A_{i_k} \cdot (\Sigma s_i A_i + \gamma \Sigma D_e) = A_{i_k} \cdot (\nu Z + Z_k + \gamma \Sigma D_e) \leq -\nu + \tau + \gamma\omega \leq -\lambda - 1$. Hence ψ extends to $\phi^{(1)}: (A \cup D)(\mathcal{I}^{\nu} \mathcal{I}_1 q^{\gamma}) \to (\tilde{A} \cup \tilde{D})(\tilde{\mathcal{I}}^{\nu} \tilde{\mathcal{I}}_1 \tilde{q}^{\gamma})$, $\phi^{(2)}, \ldots$ up to $\phi^{(r)}: (A \cup D)(\mathcal{I}^{\nu+1} q^{\gamma}) \to (\tilde{A} \cup \tilde{D})(\tilde{\mathcal{I}}^{\nu+1} \tilde{q}^{\gamma})$. Repeat this argument to get $\phi': (A \cup D)(\mathcal{I}^{\nu+\omega} q^{\gamma}) \to (A \cup D)(\tilde{\mathcal{I}}^{\nu+\omega} \tilde{q}^{\gamma})$. ϕ' will extend to ϕ if $H^1(A, {}_{n,m}\Theta) = 0$. But ${}_{n,m}\Theta$ is supported on D and is a coherent sheaf over each component of D since it is locally free. Each component of D is an open subset of the plane and hence a Stein manifold. Hence $H^1(A, {}_{n,m}\Theta) = 0$.∎

Proposition 6.16 implies that $A \cup D$ and $\tilde{A} \cup \tilde{D}$ have formally equivalent neighborhoods.

Since $c_j(\mathcal{I}^{\mu} q) \geq c_j(\mathcal{I}^{\mu}) - \omega \geq \mu - \omega$, for sufficiently large μ we may satisfy $c_j(\mathcal{I}^{\mu} q) > \sigma$, as in Lemma 6.15. Also, $c_j((\mathcal{I}^{\mu} q)^{\nu}) > \sigma$ for $\nu \geq 1$. For $\mu \geq \chi$, $\Phi_*((\mathcal{I}^{\mu} q)^{\nu})$ is coherent by the following argument. If $\mathcal{F}$ is the ideal sheaf generated by g, $\mathcal{F} = q \Pi p_i^{s_i}$ with $s_i \leq \chi$. $\Phi_*(\mathcal{F}^{\nu})$ is coherent by Lemma 5.2. Since $s_i \leq \chi \leq \mu$, $\Phi_*((\mathcal{I}^{\mu} q)^{\nu})$ is also coherent by Lemma 5.2. Thus by Lemma 6.15, we have

LEMMA 6.17. *For* $\mu > \max(\omega + \sigma, \chi)$, $\nu \geq 1$, *if* U *and* $\tilde{U}$ *are sufficiently small holomorphically convex neighborhoods of* A *and* $\tilde{A}$, $H^1(U, (\mathcal{I}^{\mu} q)^{\nu}) = H^1(\tilde{U}, (\tilde{\mathcal{I}}^{\mu} \tilde{q})^{\nu}) = 0$.

Thus given any holomorphic function f on U, we can find a holomorphic function $\tilde{f}$ on $\tilde{U}$ such that f and $\tilde{f}$ induce corresponding sections in $\mathcal{O}/(\mathcal{J}^{\mu} q)^{\nu}$ and $\mathcal{O}/(\tilde{\mathcal{J}}^{\mu} \tilde{q})^{\nu}$.

Let $\mathcal{J}$ be the ideal sheaf of A. For sufficiently large μ, $H^1(U, \mathcal{J}^2 \mathcal{J}^{\mu}) = H^1(U, \mathcal{J}\mathcal{J}^{\mu}) = H^1(U, \mathcal{J}^{\mu}) = 0$ by Lemma 6.15. $\mathcal{J}^{\mu}/\mathcal{J}^2\mathcal{J}^{\mu}$ is the sheaf of germs of sections of a vector bundle of rank 2 over A.

LEMMA 6.18. *For all sufficiently large* μ, *if* $f_o, \ldots, f_s$ *are holomorphic functions in* U *such that* $\pi(f_o), \ldots, \pi(f_s) \in \Gamma(A, \mathcal{J}^{\mu}/\mathcal{J}^2\mathcal{J}^{\mu})$ *form a basis for* $\Gamma(A, \mathcal{J}^{\mu}/\mathcal{J}^2\mathcal{J}^{\mu})$, *then the mapping* $F: U \to P^s$ *given in homogeneous coordinates by* $F(x) = (f_o(x), \ldots, f_s(x))$ *embeds a neighborhood of* A *biholomorphically into* P^s.

Proof: $f_o, \ldots, f_s$, of course, all vanish on A but they all vanish at least to the order μr_i on A_i. The ratio $f_o : \ldots : f_s$ will be well-defined and will give a holomorphic map into P^s near A if for each $x \in A$, there is a function, say f_o, whose zero set near x is A and such that f_o vanishes to exactly order μr_i on A_i and μr_j on A_j, if $x \in A_i \cap A_j$. In other words, the images of the f_i in $\Gamma(A, \mathcal{J}^{\mu}/\mathcal{J}\mathcal{J}^{\mu})$ should, as sections of a line bundle, have no common zero.

$$0 \to \mathcal{J}\mathcal{J}^{\mu}/\mathcal{J}^2\mathcal{J}^{\mu} \to \mathcal{J}^{\mu}/\mathcal{J}^2\mathcal{J}^{\mu} \to \mathcal{J}^{\mu}/\mathcal{J}\mathcal{J}^{\mu} \to 0$$

is exact. Since $H^1(U, \mathcal{J}\mathcal{J}^{\mu}) = 0$, any section of $\mathcal{J}^{\mu}/\mathcal{J}\mathcal{J}^{\mu}$ may be lifted to U and then restricted back to $\Gamma(A, \mathcal{J}^{\mu}/\mathcal{J}^2\mathcal{J}^{\mu})$. Thus

$$0 \to \Gamma(A, \mathcal{J}\mathcal{J}^{\mu}/\mathcal{J}^2\mathcal{J}^{\mu}) \to \Gamma(A, \mathcal{J}^{\mu}/\mathcal{J}^2\mathcal{J}^{\mu}) \to \Gamma(A, \mathcal{J}^{\mu}/\mathcal{J}\mathcal{J}^{\mu}) \to 0$$

is exact. We first show that for sufficiently large μ, a basis $h_o, \ldots, h_t$ for $\Gamma(A, \mathcal{J}^{\mu}/\mathcal{J}\mathcal{J}^{\mu})$ will embed A in P^t. Then, in addition, $(f_o, \ldots, f_s)$ will be a well-defined set of homogeneous coordinates and will embed A in P^s.

$c_j(\mathcal{I}^\mu/\mathcal{J}\mathcal{I}^\mu) \to \infty$ as $\mu \to \infty$ so we can find many sections over A_j all j. Sections over A_i and A_j which agree at $A_i \cap A_j$ will give sections over $A_i \cup A_j$. We thus can use the methods of the proof of Lemma 4.11. Let us first show that $(h_o, \ldots, h_t)$ is well-defined in P^t and separates points. A change in basis of $\Gamma(A, \mathcal{I}^\mu/\mathcal{J}\mathcal{I}^\mu)$ corresponds to an automorphism of P^t, so we are free to choose any basis which is convenient to the problem at hand. Let B be the line bundle over A for which $\mathcal{I}^\mu/\mathcal{J}\mathcal{I}^\mu$ is the sheaf of germs of sections. Given a regular point $a \in A_i$ we first find a section h of B over A_i such that $h(a) \neq 0$ and $h(b) = 0$ for all singular points of A lying in A_i. This puts only a finite number of conditions on the sections of the line bundle and hence by Riemann-Roch, for sufficiently large μ such an h can be found. Extend h by 0 to give an element of $\Gamma(A, \mathcal{I}^\mu/\mathcal{J}\mathcal{I}^\mu)$. For $a \in A_i \cap A_j$, find $h \in \Gamma(A_i, \mathcal{B})$, $h(a) \neq 0$ and $h' \in \Gamma(A_j, \mathcal{B})$, $h'(a) \neq 0$ such that h and h′ vanish at the other singular points. Then $h(a) = kh'(a)$ for a suitable $k \neq 0$ and this section extends to all of A. To separate points it suffices, given $a \neq b$, to find sections s_1 and s_2 such that $s_1(a) \neq 0$, $s_1(b) = 0$ $s_2(a) = 0$ and $s_2(b) \neq 0$. If a and b are in distinct A_i, the above construction already suffices. If a and b are in A_i, $s_1(b) = 0$ just imposes another vanishing condition and we can still find s_1 for large μ.

We next show that the map is locally biholomorphic on A. If a is not a singular point of A, find sections s_1 and s_2 such that $s_1(a) \neq 0$ and s_2 vanishes to exactly first order at a. If (x,y) are local coordinates near $a = (0,0)$, $A_i = \{y = 0\}$, s_1 may be locally given as the restriction of an ambient function $y^d(a_{10} + a_{11}x + \ldots)$ with $d = \mu r_i$ snd $a_{10} \neq 0$. s_2 may be given by $y^d(a_{21}x + a_{22}x + \ldots)$ with $a_{21} \neq 0$. Then s_2/s_1 is a biholomorphic map near a and, using inhomogeneous coordinates, shows that $(h_o, \ldots, h_t)$ is biholomorphic near a.

Finally, if a is a singular point and locally $A = \{xy = 0\}$, we may find sections s_1, s_2, s_3 given locally by

$s_1 = x^d y^e(a_{11}x + a_{12}x^2 + \dots)$, $a_{11} \neq 0$, $s_2 = x^d y^e(b_{21}y + b_{22}y^2 + \dots)$ $b_{21} \neq 0$ and $s_3 = x^d y^e(a_{30} + a_{31}x + \dots + b_{31}y + \dots)$, $a_{30} \neq 0$. In fact, if f_1, f_2 and f_3 are any ambient holomorphic functions representing s_1, s_2 and s_3,

$$f_1(x,y) = x^d y^e(a_{11}x + \dots) + x^{d+1}y^{e+1}k_1(x,y)$$

$$f_2(x,y) = x^d y^e(b_{21}y + \dots) + x^{d+1}y^{e+1}k_2(x,y)$$

$$f_3(x,y) = x^d y^e(a_{30} + \dots) + x^{d+1}y^{e+1}k_3(x,y) ,$$

with k_1, k_2, and k_3 holomorphic. $(f_1/f_3, f_2/f_3)$ has $(a_{11}x/a_{30}, b_{21}y/a_{30})$ as its linear terms and hence is biholomorphic in an ambient neighborhood of a.

Thus so far, $F = (f_o, \dots, f_s)$ embeds A biholomorphically and is a locally biholomorphic map in a neighborhood of the singular points. Let us next show that F is biholomorphic near any $a \in A_i$ which is a regular point of A. Any functions f_1 and f_2 representing the s_1 and s_2 derived above must be of the form

$$f_1 = y^d(a_{10} + a_{11}x + \dots) + y^{d+1}k_1(x,y) \qquad a_{10} \neq 0$$

$$f_2 = y^d(a_{21}x + \dots) + y^{d+1}k_2(x,y) \qquad a_{21} \neq 0 .$$

Let t_o be an element of $\Gamma(A, \mathcal{J}\mathcal{I}^{\mu}/\mathcal{J}^2\mathcal{I}^{\mu})$ such that as a section of the corresponding line bundle, t_o is non-zero at a. As above, such a t_o will exist for suitably large μ. Then any f_o representing t_o must be given locally by

$$f_o = y^{d+1}(a_{00} + a_{10}x + \dots) + y^{d+2}k_o(x,y) , \qquad a_{00} \neq 0 .$$

Then $(f_2/f_1, f_0/f_1)$ has $(a_{21}x/a_{10} + cy, a_{00}y/a_{10})$ as its linear terms and hence is biholomorphic in a neighborhood of a.

Thus $F: U \to (f_o, \ldots, f_s)$ is one-to-one on A and biholomorphic in a neighborhood of each point of A. Since A is compact, F is one-to-one and thus biholomorphic in a sufficiently small neighborhood of A.∎

If $\tilde{f}_o, \ldots, \tilde{f}_s$ are holomorphic functions in a neighborhood of $\tilde{A}$ whose images in $\Gamma(\tilde{A}, \tilde{\mathcal{J}}^{\mu}/\mathcal{J}^2\tilde{\mathcal{J}}^{\mu})$ correspond to $\pi(f_o), \ldots, \pi(f_s)$, then Lemma 6.18 shows that $(\tilde{f}_o, \ldots, \tilde{f}_s)$ embeds some neighborhood $\tilde{U}$ of $\tilde{A}$ into $\mathbf{P}^s$.

For a general $\mathbf{C}^w$, let T be its tangent bundle. Suppose that X is a not necessarily closed complex submanifold of $\mathbf{C}^w$. Let ${}_X T$ be the tangent bundle to X. If $x \in X$, an affine subspace N(x) of $\mathbf{C}^w$ through x is said to be normal to X at x if the tangent space of N(x) at x is complementary to ${}_X T_x$ in T_x. A holomorphic choice of N(x) is called a normal field along X. Recall Proposition 6.14 and its proof. Let S be the $(n-2) \times n$ matrix (t_{ej}), $1 \leq e \leq n-2$, $1 \leq j \leq n$ which explicitly gives the $s_1, \ldots, s_{n-2} \in \Gamma(\Delta, \mathcal{T})$. For each $z_o \in Y - V(g) = \Phi(U) - \Phi(A \cup D)$, the affine subspaces $N_o(z_o) = \{z \in \mathbf{C}^n \mid z = z_o + vS$, where v is some $1 \times (n-2)$ matrix$\}$ form a normal field to $Y - \Phi(A \cup D)$. Let $\Phi = (f_{s+1}, \ldots, f_{s+n})$. Then

$$G: z \to ((f_o(z), \ldots, f_s(z)), f_{s+1}(z), \ldots, f_{s+n}(z))$$

maps a neighborhood U of A biholomorphically onto a non-closed submanifold of $\mathbf{P}^s \times \mathbf{C}^n$. Let $\pi: \mathbf{P}^s \times \mathbf{C}^n \to \mathbf{C}^n$ be the projection map. For $z \in G(U - (A \cup D))$, let $N(z) = \mathbf{P}^s \times N_o(\pi(z))$. Then the tangent space to N(z) at z is complementary to the tangent space of G(U) at a. If W is a coordinate patch in $\mathbf{P}^s$, we claim that N(z) is an affine subspace near z in $W \times \mathbf{C}^n$. But this is trivial since $N(z) \cap (W \times \mathbf{C}^n) = W \times N_o(\pi(z))$ which is locally an affine subspace. Thus the N(z) form a normal field to $G(U - (A \cup D))$ in $W \times \mathbf{C}^n$.

Our final steps in the proof of Theorem 6.13 will be as follows. From Proposition 6.16, $A \cup D$ and $\tilde{A} \cup \tilde{D}$ have formally equivalent neighborhoods. Choose μ sufficiently large so that the conclusions of Lemma 6.17 and Lemma 6.18 hold. For any $\nu \geq 1$ and any $\tilde{f}_o, \ldots, \tilde{f}_s$, $\tilde{f}_{s+1}, \ldots, \tilde{f}_{s+n}$ holomorphic near $\tilde{A}$ and having images in $\mathcal{O}/(\tilde{\mathcal{J}}^{\mu}\tilde{q})^{\nu}$ corresponding to the images of $f_o, \ldots, f_s, f_{s+1}, \ldots, f_n$ in $\mathcal{O}/(\mathcal{J}^{\mu} q)^{\nu}$, $\tilde{G} = ((\tilde{f}_o, \ldots, \tilde{f}_s), \tilde{f}_{s+1}, \ldots, \tilde{f}_{s+n})$ embeds a neighborhood $\tilde{U}$ of $\tilde{A}$ biholomorphically in $\mathbf{P}^s \times \mathbf{C}^n$. We shall identify U and $\tilde{U}$ with their images in $\mathbf{P}^s \times \mathbf{C}^n$. $A \cup D$ and $\tilde{A} \cup \tilde{D}$ then coincide. Namely, on A and $\tilde{A}$, f_ℓ and $\tilde{f}_\ell$, $\ell > s$, vanish and the embedding in $\mathbf{P}^s$ is determined by $\mathcal{J}^{\mu}/\mathcal{J}^2\mathcal{J}^{\mu}$ and $\tilde{\mathcal{J}}^{\mu}/\tilde{\mathcal{J}}^2\tilde{\mathcal{J}}^{\mu}$. For $z \in A \cup D$, letting $\tilde{z} = \phi(z)$, $\tilde{f}_\ell(\tilde{z}) = f_\ell(z)$, all ℓ. Thus on $D-A$ and $\tilde{D}-\tilde{A}$, where $(f_o, \ldots, f_s) \neq (0, \ldots, 0)$, $G(z) = \tilde{G}(\tilde{z})$. For each $z \in U - (A \cup D)$, we have $N(z)$ defined. We shall show that for sufficiently large ν, the map $z \to N(z) \cap \tilde{U}$, $z \notin A \cup D$, and $z \to z$, $z \in A \cup D$, is well-defined and establishes a biholomorphic map between sufficiently small neighborhoods of A and $\tilde{A}$.

We now wish to compare the defining equations for U and $\tilde{U}$ in $\mathbf{P}^s \times \mathbf{C}^n$. Let (x,y) be coordinates on U near a point $(0,0) \in A \cap D$, $A = \{y = 0\}$, $D = \{x = 0\}$. The other cases, regular points in $A \cup D$ and singular points in A, can be handled in an entirely similar manner. $G = ((f_o, \ldots, f_s), f_{s+1}, \ldots, f_{s+n})$ is given locally by $((y^d h_o(x,y), \ldots, y^d h_s), y^d h_{s+1}, \ldots, y^d h_{s+n})$. By making a non-singular linear change among the $f_o, \ldots, f_s$ we may assume that $h_o(0,0) = 1$, $h_1(x,y)$ has x as its linear terms and $h_2(x,y)$ has y as its linear terms. We may choose coordinates $(\tilde{x}, \tilde{y})$ for $\tilde{U}$ so that near $(0,0)$, $\psi : \mathcal{O}/(\tilde{\mathcal{J}}^{\mu}\tilde{q})^{\nu} \to \mathcal{O}/(\mathcal{J}^{\mu} q)^{\nu}$ is induced by $\tilde{y} = y$, $\tilde{x} = x$. Then $\tilde{A} = \{\tilde{y} = 0\}$ and $\tilde{D} = \{\tilde{x} = 0\}$. Then

$$\tilde{f}_k(x,y) = f_k(x,y) + x^{\nu} y^{\nu} g_k(x,y) \tag{6.6}$$

for holomorphic $g_k(x,y)$. Now let $z_1,\ldots,z_s, z_{s+1},\ldots,z_{s+n}$ be inhomogeneous coordinates for $P^s \times C^n$. Then G is given by $z_i = f_i/f_o$, $1 \leq i \leq s$, $z_j = f_j$, $s+1 \leq j \leq s+n$. $\tilde{G}$ is given similarly. $z_1 = x +$ (quadratic terms) and $z_2 = y +$ (quadratic terms). Assuming that $\nu \geq d+2$, we also have that $z_1 = \tilde{x} +$ (quadratic terms in $\tilde{x}$ and $\tilde{y}$), $z_2 = \tilde{y} +$ (quadratic terms in $\tilde{x}$ and $\tilde{y}$). Thus to get defining equations for U we solve for x and y in terms of z_1 and z_2 and substitute these expressions in our equations for z_i, $i \geq 3$. Similarly, we can solve for $\tilde{x}$ and $\tilde{y}$ in terms of z_1 and z_2 and get equations for $\tilde{U}$. We can in particular think of $x,y,\tilde{x}$, and $\tilde{y}$ as functions of z_1 and z_2 and regard z_1 and z_2 as local coordinates for both U and $\tilde{U}$. xy and $\tilde{x}\tilde{y}$ then generate the ideas of A U D and $\tilde{A}$ U $\tilde{D}$ respectively.

From (6.6) for $1 \leq i \leq s$, and especially for $i = 1,2$,

$$z_i = \frac{\tilde{h}_i(\tilde{x},\tilde{y})}{\tilde{h}_o(\tilde{x},\tilde{y})} = \frac{h_i(x,y)}{h_o(x,y)} = \frac{\tilde{h}_i(x,y) - x^\nu y^{\nu-d} g_i(x,y)}{\tilde{h}_o(x,y) - x^\nu y^{\nu-d} g_o(x,y)}$$

$$(6.7) \qquad = \frac{\tilde{h}_i(x,y)}{\tilde{h}_o(x,y)} + x^\nu y^{\nu-d}\, k_i(x,y) \;.$$

By the inverse mapping theorem, there is a pair (H_1, H_2) of holomorphic functions such that $(H_1(\tilde{h}_1(\tilde{x},\tilde{y})/\tilde{h}_o), H_2(\tilde{h}_2/\tilde{h}_o)) = (\tilde{x},\tilde{y})$. Applying (H_1,H_2) to (6.7), we get $\tilde{x} = x + x^\nu y^{\nu-d}\, u(x,y)$ and $\tilde{y} = y + x^\nu y^{\nu-d}\, v(x,y)$ for holomorphic u and v. The defining equations for $\tilde{U}$ are $z_i = \tilde{f}_i = \tilde{f}_i(\tilde{x},\tilde{y})/\tilde{f}_o(\tilde{x},\tilde{y}) = \tilde{h}_i(\tilde{x},\tilde{y})/\tilde{h}_o(\tilde{x},\tilde{y})$, $3 \leq i \leq s$, $z_j = \tilde{f}_j(\tilde{x},\tilde{y})$, $s+1 \leq j \leq s+n$. Modulo the ideal generated by $x^\nu y^{\nu-d}$, $\tilde{h}_i(\tilde{x},\tilde{y})/\tilde{h}_o(\tilde{x},\tilde{y}) \equiv \tilde{h}_i(x,y)/\tilde{h}_o(x,y)$ since we have just shown that $x \equiv \tilde{x}$ and $y \equiv \tilde{y}$. From (6.7), for $3 \leq i \leq s$, $\tilde{h}_i(x,y)/\tilde{h}_o(x,y) \equiv h_i(x,y)/h_o(x,y)$

modulo $x^{\nu}y^{\nu-d}$. Similarly for $s+1 \leq j \leq s+n$. Let $\eta = \max_{\mu} r_i$ so that η is the maximum of the d's. Let $\sigma = \nu - \eta$. We have shown the following. Near each point of $A \cup D$ and $\tilde{A} \cup \tilde{D}$ there are coordinates for $W \times C^n \subset P^s \times C^n$ with U given by

$$z_i = m_i(z_1, z_2) \qquad 3 \leq i \leq s+n$$

and $\tilde{U}$ given by

$$z_i = n_i(z_1, z_2) \qquad 3 \leq i \leq s+n$$

such that if $I(z_1,z_2)$ $(= xy)$ generates the ideal of $A \cup D$ in the (z_1,z_2) coordinate system, then

$$n_i(z_1, z_2) - m_i(z_1, z_2) = I^{\sigma}(z_1, z_2)\alpha_i(z_1, z_2)$$

for some holomorphic function α_i. We may make σ arbitrarily large by suitable choice of the $\tilde{f}_i$. This will change the n_i and the α_i but will not change the m_i and I.

In $W \times C^n$, the tangent space to U at $\hat{z} = (\hat{z}_1, \hat{z}_2, \ldots, \hat{z}_{s+n})$ is given by

$$z_j = \hat{z}_j + \frac{\partial m_j}{\partial z_j}(z_1 - \hat{z}_1) + \frac{\partial m_j}{\partial z_2}(z_2 - \hat{z}_2) \qquad 3 \leq j \leq s+n\,.$$

Let

$$z_j^* = z_j - \hat{z}_j - \frac{\partial m_j}{\partial z_1}(z_1 - \hat{z}_1) - \frac{\partial m_j}{\partial z_2}(z_2 - \hat{z}_2)\,. \tag{6.8}$$

Recall that $N(\hat{z}) = W \times N_o(\pi(\hat{z}))$ is a normal affine subspace to $\hat{z}$ if $\hat{z} \notin A \cup D$. $N_o(\pi(\hat{z}))$ is given by $\{z_j = \hat{z}_j + (vS)_j\}$, $s+1 \leq j \leq s+n$, where v is a $1 \times (n-2)$ matrix and S is an $(n-2) \times n$ matrix with constant entries. $(vS)_j$ is the jth entry in the $1 \times n$ matrix vS.

Letting $\hat{z}$ be the origin, we have the following simple situation from linear algebra. Suppose $\hat{z} \notin A \cup D$. Regard the z_j^*, $3 \leq j \leq s+n$, as linear functions. Then the tangent space to U at $\hat{z}$ is the common zeros of the z_j^*. The subspace $N(\hat{z})$ is complementary to the tangent space. Thus the common zeros of the restrictions of the z_j^* to $N(\hat{z})$ is just $\hat{z} = 0$. So the z_j^* are a basis for the linear functionals on $N(\hat{z})$. $z_1 - \hat{z}_1$ and $z_2 - \hat{z}_2$ are linear functionals on $N(\hat{z})$. Expressing these linear functionals in terms of the z_j^* gives defining equations of the following form for $N(\hat{z})$.

$$z_1^* = z_1 - \hat{z}_1 = \sum_{j=3}^{s+n} b_{1j}(\hat{z}_1, \hat{z}_2) z_j^*$$

(6.9)

$$z_2^* = z_2 - \hat{z}_2 = \sum_{j=3}^{s+n} b_{2j}(\hat{z}_1, \hat{z}_2) z_j^*$$

Using Kramer's rule, we see that the b_{ij} are rational functions of the entries in S and the $\frac{\partial m_j}{\partial z_1}$ and $\frac{\partial m_j}{\partial z_2}$. The b_{ij} are holomorphic for $(\hat{z}_1, \hat{z}_2) \notin A \cup D$ and may have poles for $(\hat{z}_1, \hat{z}_2) \in A \cup D$.

Thus all that remains to be proven in the proof of Theorem 6.13 is the following.

THEOREM 6.19. *Let* $\rho \geq 1$ *be chosen large enough so that* $I^\rho(z_1, z_2) b_{ij}(z_1, z_2)$, $i = 1,2$, $j = 3, 4, \ldots, s+n$ *are holomorphic on* U *and vanish on* $A \cup D$. *Let* σ *then be chosen large enough so that* $\sigma - \rho \geq 1$ *and* $I^{\sigma-1} b_{ij}$ *and* $I^{\sigma-\rho} b_{ij}$, $i = 1,2$, $j = 3,4,\ldots,s+n$ *are holomorphic on* U *and vanish on* $A \cup D$. *Then there is a neighborhood of* A U D *where there are uniquely determined functions* $\beta_1(z_1,z_2)$ *and* $\beta_2(z_1,z_2)$ *such that under the map* $(z_1,z_2) \to (\tilde{z}_1,\tilde{z}_2) = (z_1 + I^{\rho}\beta_1,\ z_2 + I^{\rho}\beta_2)$, *the normal* $N(z) = N(z_1,z_2,\ldots,m_j(z_1,z_2),\ldots)$ *meets* $\tilde{U}$ *at* $(\tilde{z}_1,\tilde{z}_2,\ldots,n_j(\tilde{z}_1,\tilde{z}_2),\ldots)$. β_1 *and* β_2 *are holomorphic. Hence the map* $(z_1,z_2) \to (\tilde{z}_1,\tilde{z}_2)$ *is locally biholomorphic. Since the map* $\psi : U \to \tilde{U}$ *given locally by* $(z_1,z_2) \to (\tilde{z}_1,\tilde{z}_2)$ *is independent of the coordinate patch for* $\mathbf{P}^s \times \mathbf{C}^n$ *and maps* A U D *identically to* $\tilde{A} \cup \tilde{D}$, ψ *establishes a biholomorphic map between some neighborhood of* A U D *and some neighborhood of* $\tilde{A} \cup \tilde{D}$.

Proof: We are going to use the implicit mapping theorem. N(z) is given by (6.8) and (6.9).

$$\tilde{z}_i = z_i = \sum_{j=3}^{s+n} b_{ij}(z_1,z_2)\,(\tilde{z}_j - z_j - \frac{\partial m_j}{\partial z_1}(\tilde{z}_1 - z_1) - \frac{\partial m_j}{\partial z_2}(\tilde{z}_2 - z_2)),$$

$i = 1,2$. $z_j = m_i(z_1,z_2)$ and $\tilde{U}$ meets N(z) at $\tilde{z}_j = n_j(\tilde{z}_1,\tilde{z}_2)$. Hence

$$\tilde{z}_i - z_i = \sum_{j=3}^{s+n} b_{ij}(z_1,z_2)(n_j(\tilde{z}_1,\tilde{z}_2) - m_j(z_1,z_2) - \frac{\partial m_j}{\partial z_1}(\tilde{z}_1 - z_2) -$$

$$\frac{\partial m_j}{\partial z_2}(\tilde{z}_2 - z_2)), \qquad i = 1,2.$$

This gives us two equations which implicitly define the map $(z_1,z_2) \to (\tilde{z}_1,\tilde{z}_2)$. We may rewrite them as follows.

$(\tilde{z}_1,\tilde{z}_2) = (z_1 + I^{\rho}(z_1,z_2)\beta_1, z_2 + I^{\rho}(z_1,z_2)\beta_2)$. $n_j - m_j = I^{\sigma}\alpha_j$. Thus for $i = 1,2$,

$$0 = H_i(z_1,z_2,\beta_1,\beta_2) =$$

$$\beta_i - \sum_{j=3}^{s+n} b_{ij}(z_1,z_2)I^{-\rho}(z_1,z_2)[n_j(z_1,z_2) - m_j(z_1,z_2)$$

(6.10)

$$+ m_j(\tilde{z}_1,\tilde{z}_2) - m_j(z_1,z_2) - \frac{\partial m_j}{\partial z_1}(\tilde{z}_1 - z_1) - \frac{\partial m_j}{\partial z_2}(\tilde{z}_2 - z_2)$$

$$+ n_j(\tilde{z}_1,\tilde{z}_2) - m_j(\tilde{z}_1,\tilde{z}_2) - n_j(z_1,z_2) + m_j(z_1,z_2)]$$

$$= \beta_1 - \sum_{j=3}^{s+n} b_{ij}(z_1,z_2)[I^{\sigma-\rho}(z_1,z_2)\alpha_j(z_1,z_2)$$

$$+ I^{\rho}(z_1,z_2)\gamma_j(z_1,z_2,\beta_1,\beta_2) + I^{\sigma-1}(z_1,z_2)\lambda_j(z_1,z_2,\beta_1,\beta_2)].$$

where

$$\gamma_j(z_1,z_2,\beta_1,\beta_2) = I^{-2\rho}(z_1,z_2)[m_j(z_1+I^{\rho}(z_1,z_2)\beta_1,\ z_2+I^{\rho}\beta_2)$$

$$- m_j(z_1,z_2) - \frac{\partial m_j}{\partial z_1} I^{\rho}\beta_1 - \frac{\partial m_j}{\partial z_2} I^{\rho}\beta_2] \qquad \text{and}$$

$$\lambda_j(z_1,z_2,\beta_1,\beta_2) = I^{1-\rho-\sigma}(z_1,z_2)[n_j(z_1+I^{\rho}\beta_1,z_2+I^{\rho}\beta_2)$$

$$- m_j(z_1+I^{\rho}\beta_1,\ z_2+I^{\rho}\beta_2) - n_j(z_1,z_2) + m_j(z_1,z_2)]$$

$$= I^{1-\rho-\sigma}(z_1,z_2)[I^{\sigma}(z_1+I^{\rho}\beta_1,\ z_2+I^{\rho}\beta_2)\,\alpha_j(z_1+I^{\rho}\beta_1,z_2+I^{\rho}\beta_2)$$

$$- I^{\sigma}(z_1,z_2)\,\alpha_j(z_1,z_2)].$$

In (6.10), α_j is holomorphic. We claim that γ_j and λ_j are also holomorphic. For γ_j, $m_j + \frac{\partial m_j}{\partial z_1} I^{\rho}\beta_1 + \frac{\partial m_j}{\partial z_2} I^{\rho}\beta_2$ are the first two terms in the power series expansion of $m_j(z_1 + I^\rho\beta_1, z_2 + I^\rho\beta_2)$ so that $I^{2\rho}$ divides $m_j(z_1 + I^\rho\beta_1, z_2 + I^\rho\beta_2) -$

$[m_j + \frac{\partial m_j}{\partial z_1} I^\rho\beta_1 + \frac{\partial m_j}{\partial z_2} I^\rho\beta_2]$ and γ_j is holomorphic.

We may write λ_j as follows. $\lambda_j(z_1,z_2,\beta_1,\beta_2) =$

$$I^{1-\rho-\sigma}(z_1,z_2)\,[I^\sigma(z_1 + I^\rho\beta_1, z_2 + I^\rho\beta_2)\,\alpha_j(z_1 + I^\rho\beta_1, z_2 + I^\rho\beta_2)$$

$$- I^\sigma(z_1,z_2)\,\alpha_j(z_1,z_2)]$$

$$= I^{1-\rho-\sigma}(z_1,z_2)[I^\sigma(z_1,z_2)\,\{\alpha_j(z_1 + I^\rho\beta_1, z_2 + I^\rho\beta_2) - \alpha_j(z_1,z_2)\}$$

$$+\{I^\sigma(z_1 + I^\rho\beta_1, z_2 + I^\rho\beta_2) - I^\sigma(z_1,z_2)\}\,\alpha_j(z_1 + I^\rho\beta_1, z_2 + I^\rho\beta_2)\}\,.$$

Using a power series expansion, we see that I^ρ divides $\alpha_j(z_1 + I^\rho\beta_1, z_2 + I^\rho\beta_2) - \alpha_j(z_1,z_2)$. Thus to show that λ_j is holomorphic, we have only left to show that $I^{\rho+\sigma-1}$ divides $I^\sigma(z_1 + I^\rho\beta_1, z_2 + I^\rho\beta_2) -$ $I^\sigma(z_1,z_2)$. $I(z_1 + I^\rho\beta_1, z_2 + I^\rho\beta_2) = I(z_1,z_2) + I^\rho(z_1,z_2)\,J(z_1,z_2)$ with $J(z_1,z_2)$ holomorphic. Hence, using the binomial theorem and $\rho \geq 1$, we see that $I^{\rho+\sigma-1}$ divides $I^\sigma(z_1 + I^\rho\beta_1, z_2 + i^\rho\beta_2) - I^\sigma(z_1,z_2)$.

Now return to the equations (6.10), which are analytic equations in z_1,z_2 and β_1,β_2. Since $I^{\sigma-\rho}$, I^ρ and $I^{\sigma-1}$ all vanish on $A \cup D$, on $A \cup D$, $(\frac{\partial H_i}{\partial \beta_j}) = (\begin{smallmatrix} 1 & 0 \\ 0 & 1 \end{smallmatrix})$. Then by the implicit mapping theorem (Theorem I.B.5 of G & R), in a neighborhood of each point of $A \cup D$, there exist unique holomorphic functions $\beta_1(z_1,z_2)$ and $\beta_2(z_1,z_2)$ such that (6.10) is satisfied if and only if $\beta_1 = \beta_1(z_1,z_2)$ and $\beta_2 = \beta_2(z_1,z_2)$.▮

This completes the proof of Theorem 6.19 and thus the proof of Theorem 6.13.▮

Theorems 6.9 and 6.13 together give

THEOREM 6.20. *Let* A *and* $\tilde{A}$ *be 1-dimensional analytic subsets of the 2-dimensional manifolds* M *and* $\tilde{M}$ *respectively. Suppose that* A *is exceptional in* M *and its irreducible components are non-singular, cross transversely and no three components meet at a point. Given any isomorphism* $\phi: A(\mathcal{I}^{\nu}) \to \tilde{A}(\tilde{\mathcal{I}}^{\nu})$ *with* $\nu \geq \lambda + \tau + 1$, *there exist neighborhoods* U *and* $\tilde{U}$ *of* A *and* $\tilde{A}$ *respectively and a biholomorphic map* $\psi: U \to \tilde{U}$ *such that* ϕ *and* ψ *induce the same map between the reduced spaces* A *and* $\tilde{A}$.

One consequence of this chapter's massive machinery is the following.

PROPOSITION 6.21. *Let* $\tilde{p} \in V$ *be a normal 2-dimensional singularity with a resolution* $\pi: \tilde{M} \to V$, $\tilde{A} = \pi^{-1}(\tilde{p})$. *If* $\tilde{A}$ *is a non-singular projective line and* $\tilde{A} \cdot \tilde{A} = -k$, $k \geq 2$, *then* $\tilde{p}$ *is equivalent to the singularity obtained by blowing down* A *in* $M(k)$.

Proof: Let $\phi_0: A \to \tilde{A}$ be an isomorphism of the reduced spaces. Let $m_0 = p = \mathcal{I}d(A)$ and $m_1 = p^2$. Then $\mathcal{A}ut(m_1, m_0) \approx {}_p\Theta \otimes m_0/m_1$. $c({}_p\Theta) = 2$ and $c(m_0/m_1) = 1$ so $H^1(A, \mathcal{A}ut(m_1, m_0)) = 0$. $\mathcal{A}n(m_1, m_0) \approx \mathcal{O}^*$. $H^1(A, \mathcal{O}^*) \approx Z$. From Theorem 6.5, we have that

$$(6.11) \qquad * \to H^1(A, \mathcal{A}ut(m_1 : m_0)) \to Z$$

is an exact sequence of pointed sets. We wish to apply Theorem 6.6. By carefully examining the definitions in this case, one verifies that the image of $[\phi_0]$ in Z corresponds to the difference of the Chern classes of the normal bundles of A and $\tilde{A}$. Since (6.11) is exact, $[\phi_0] = *$ and ϕ_0 extends to ϕ_1. $m_2 = p^3$ and $\mathcal{A}n(m_2, m_1) = 1$ and $\mathcal{A}ut(m_2, m_1) \approx \Theta \otimes m_1/m_2$.

$$0 \to {}_p\Theta \to \Theta \to \mathfrak{N} \to 0$$

is exact so

$$0 \to {}_p\Theta \otimes m/n \to \Theta \otimes m/n \to \mathfrak{N} \otimes m/n \to 0$$

is also exact. $c({}_p\Theta \otimes m/n) = 2 + k$. $c(\mathfrak{N} \otimes m/n) = -k + k = 0$. Hence $H^1(A, \mathcal{A}ut(m_2, m_1)) = 0$. Hence ϕ_1 extends to ϕ_2. Similar calculations show that A and $\tilde{A}$ have formally equivalent neighborhoods. Hence, this proposition follows from Theorem 6.13. ∎

CHAPTER VII

THE LOCAL RING STRUCTURE

THEOREM 7.1. *Let* p *be a normal 2-dimensional singularity. Let* $\mathcal{O}_p$ *be the local ring of germs of holomorphic functions at* p *and let* m *be the maximal ideal in* $\mathcal{O}_p$. *Let* A *be the exceptional set in a resolution of a neighborhood of* p *such that the irreducible components of* A *are non-singular, intersect transversely, and no three meet at a point. There is an integer* λ, *depending only on the genera of the irreducible components of* A *and their intersection matrix, such that if* $\tilde{p}$ *is another 2-dimensional normal singularity such that* $\mathcal{O}_p/m^\lambda$ *and* $\mathcal{O}_{\tilde{p}}/\tilde{m}^\lambda$ *are isomorphic as* $\mathbb{C}$*-algebras, then* p *and* $\tilde{p}$ *have biholomorphic neighborhoods.*

The proof of Theorem 7.1 will take almost the entire chapter. We will work primarily with the $\mathcal{J}^\mu$ of section VI rather than with m. Analytic spaces will be reduced unless otherwise specified.

LEMMA 7.2. (Riemann) *Let* A *be a 1-dimensional compact analytic space with non-singular irreducible components* $\{A_i\}$ *which cross transversely and such that no three meet at a point. Let* ω *be the number of singular points in* A *and let* g_i *be the genus of* A_i. *Let* L *be a line bundle over* A *and* $c_i(L) = c(L \mid A_i)$. *If for all* i, $c_i(L) \geq (2g_i - 2) + 1 + \omega$, *then*

$$\dim \Gamma(A, \mathcal{L}) = -\omega + \sum_i (c_i(L) - g_i + 1). \tag{7.1}$$

Proof: A section of L over all of A consists of sections over the individual A_i which agree at the singular points. Let Q be that subspace of $\Gamma(A,\mathcal{L})$ of sections which vanish at the singular points. Q is isomorphic to the direct sum of the spaces of sections over the A_i which vanish at the singular points. Hence, by the Riemann-Roch theorem for Riemann surfaces, $\dim Q = -2\omega + \Sigma_i (c_i(L) - g_i + 1)$ since there are a total of 2ω vanishing conditions with no more than ω vanishing conditions on any A_i. To generate $\Gamma(A,\mathcal{L})$ we need Q and sections with arbitrary values at each singular point. A section of L with given values at the singular points is easily constructed by first constructing the section on each A_i. There are ω singular points, so (7.1) is correct. ∎

LEMMA 7.3. *Let* A *be as in Lemma 7.2. Let* L *and* M *be line bundles over* A *with* $c_i(L) \geq 4g_i + 2\omega$ *and* $c_i(M) \geq 4g_i + 2\omega$. *Then the canonical map* $\Gamma(A,\mathcal{L}) \otimes_C \Gamma(A,\mathcal{M}) \to \Gamma(A,\mathcal{L} \otimes \mathcal{M})$ *is surjective.*

Proof: It follows from our given estimate on $c_i(L)$, Riemann-Roch, and the usual patching construction for getting sections over A, given sections over the A_i, that there is an $f \in \Gamma(A,\mathcal{L})$ such that f does not vanish at any singular point and f does not vanish identically on any A_i. $f \otimes \Gamma(A,\mathcal{M}) \to \Gamma(A,\mathcal{L} \otimes \mathcal{M})$ is then an injection and hence by Lemma 7.2 has an image with dimension $-\omega + \Sigma (c_i(M) - g_i + 1)$. Let $\{P_{ij}\}$ be the points of A_i where f vanishes and let a_{ij} be the order of the zero of f at P_{ij}. By our choice of f, no P_{ij} is a singular point. Hence by [Gu, p. 103]

$$\sum_{i,j} a_{ij} = \sum_i (\sum_j a_{ij}) = \sum_i c_i(L) .$$

By Lemma 7.2, $\dim \Gamma(A,\mathcal{L} \otimes \mathcal{M}) = -\omega + \Sigma (c_i(L) + c_i(M) - g_i + 1)$. Thus $\sum_{i,j} a_{ij}$ is also the codimension in $\Gamma(A,\mathcal{L} \otimes \mathcal{M})$ of the image of $f \otimes L(A,\mathcal{M})$. Since for fixed i, $c_i(L \otimes M) - \sum_j a_{ij} = c_i(L \otimes M) - c_i(L) \geq (2g_i - 2) + 1 + \omega$, the subspace Q of $\Gamma(A,\mathcal{L} \otimes \mathcal{M})$ whose elements vanish to order at least a_{ij} at the P_{ij} also has codimension $\sum_{i,j} a_{ij}$.

Thus the image of $f \times \Gamma(A, \mathfrak{M})$ is precisely Q. Hence to prove the lemma, it suffices to find elements of the image which vanish at the P_{ij} to exactly the given orders b_{ij}, $0 \le b_{ij} \le a_{ij}$. Given such a set of b_{ij}, choose c_{ij} so that $0 \le c_{ij} \le b_{ij}$, $c_{ij} = 0$ if $\sum_j b_{ij} \le c_i(L) - (2g_i - 2) - 1 - \omega - 1$ and $\sum_j c_{ij} = (2g_i - 2) + 1 + \omega + 1$ if $\sum_j b_{ij} > c_i(L) - (2g_i - 2) - 1 - \omega - 1$. Let $d_{ij} = b_{ij} - c_{ij}$. Then we can find $h \in \Gamma(A, \mathfrak{L})$ and $g \in \Gamma(A, \mathfrak{M})$ such that h vanishes to order exactly d_{ij} at the P_{ij} and g vanishes to order exactly c_{ij} at the P_{ij}. Namely, $\sum_j c_{ij} \le (2g_i - 2) + 1 + \omega + 1 \le c_i(M) - (2g_i - 2) - 1 - \omega - 1$ so g exists by Lemma 7.2. $\sum_j b_{ij} \le c_i(L)$, so $\sum_j d_{ij} \le c_i(L) - (2g_i - 2) - 1 - \omega - 1$ and h exists by Lemma 7.2. $h \otimes g$ then vishes to order exactly b_{ij} at P_{ij} as desired. ∎

We now return to our usual situation so that A is as in Theorem 7.1. Recall the definitions for $\mathfrak{I}$ and for $\mathfrak{I}_k$ given just before Theorem 6.9.

LEMMA 7.4. *For all sufficiently large $\mu \ge \mu_0$, μ_0 depending only on the intersection matrix and genera of the* A_i, *the canonical map* $\Gamma(A, (\mathfrak{I}^{\mu}/\mathfrak{I}^{2\mu})^{\beta}) \otimes_C \Gamma(A, (\mathfrak{I}^{\mu}/\mathfrak{I}^{2\mu})^{\gamma}) \to \Gamma(A, (\mathfrak{I}^{\mu}/\mathfrak{I}^{2\mu})^{\beta+\gamma})$ *is surjective for all* $\beta, \gamma \ge 1$.

Proof: Consider the following exact sheaf sequences:

(7.2)

$$0 \to \mathfrak{I}_1\mathfrak{I}^{\mu\beta}/\mathfrak{I}^{\mu\beta+\mu} \to \mathfrak{I}^{\mu\beta}/\mathfrak{I}^{\mu\beta+\mu} \to \mathfrak{I}^{\mu\beta}/\mathfrak{I}_1\mathfrak{I}^{\mu\beta} \to 0$$

$$0 \to \mathfrak{I}_2\mathfrak{I}^{\mu\beta}/\mathfrak{I}^{\mu\beta+\mu} \to \mathfrak{I}_1\mathfrak{I}^{\mu\beta}/\mathfrak{I}^{\mu\beta+\mu} \to \mathfrak{I}_1\mathfrak{I}^{\mu\beta}/\mathfrak{I}_2\mathfrak{I}^{\mu\beta} \to 0$$

$$\vdots$$

$$0 \to \mathfrak{I}^{\mu\beta+1}/\mathfrak{I}^{\mu\beta+\mu} \to \mathfrak{I}_{r-1}\mathfrak{I}^{\mu\beta}/\mathfrak{I}^{\mu\beta+\mu} \to \mathfrak{I}_{r-1}\mathfrak{I}^{\mu\beta}/\mathfrak{I}^{\mu\beta+1} \to 0$$

$$\vdots$$

$$0 \to \mathfrak{I}_k\mathfrak{I}^{\mu\beta+\nu}/\mathfrak{I}^{\mu\beta+\mu} \to \mathfrak{I}_{k-1}\mathfrak{I}^{\mu\beta+\nu}/\mathfrak{I}^{\mu\beta+\mu} \to \mathfrak{I}_{k-1}\mathfrak{I}^{\mu\beta+\nu}/\mathfrak{I}_k\mathfrak{I}^{\mu\beta+\nu} \to 0$$

$$\vdots$$

$$0 \to \mathfrak{I}_{r-1}\mathfrak{I}^{\mu\beta+\mu-1}/\mathfrak{I}^{\mu\beta+\mu} \to \mathfrak{I}_{r-2}\mathfrak{I}^{\mu\beta+\mu-1}/\mathfrak{I}^{\mu\beta+\mu} \to \mathfrak{I}_{r-2}\mathfrak{I}^{\mu\beta+\mu-1}/\mathfrak{I}_{r-1}\mathfrak{I}^{\mu\beta+\mu-1} \to 0$$

The short exact sequences of (7.2) successively decompose $\mathcal{I}^{\mu\beta}/\mathcal{I}^{\mu\beta+\mu}$ and the other middle terms into a subsheaf and a quotient sheaf which is the sheaf of germs of sections of a line bundle over some A_i. In addition, $\mathcal{I}_{r-1}\mathcal{I}^{\mu\beta-\mu-1}/\mathcal{I}^{\mu\beta+\mu}$ is the sheaf of germs of sections of a line bundle. With σ as in Lemma 6.15, choose μ sufficiently large so that $-A_i \cdot Z_k + \mu > \sigma$ for all A_i. Then $c_i(\mathcal{I}_k\mathcal{I}^{\mu\beta+\nu}) = -A_i \cdot Z_k - (\mu\beta+\nu)(A_i \cdot Z) \geq -A_i \cdot Z_k + \mu$ so $H^1(A, \mathcal{I}_k\mathcal{I}^{\mu\beta+\nu}) = 0$ by Lemma 6.15. Hence $\Gamma(A, \mathcal{I}_{k-1}\mathcal{I}^{\mu\beta+\nu}) \to \Gamma(A, \mathcal{I}_{k-1}\mathcal{I}^{\mu\beta+\nu}/\mathcal{I}_k\mathcal{I}^{\mu\beta+\nu})$ is surjective. Hence the corresponding sequences of sections in (7.2) are exact. There is a similar construction for $(\mathcal{I}^{\mu}/\mathcal{I}^{2\mu})^{\beta+\gamma}$.

Choose μ sufficiently large, as in Lemma 7.3, so that $\Gamma(A, \mathcal{I}^{\mu\beta}/\mathcal{I}_1\mathcal{I}^{\mu\beta}) \otimes \Gamma(A, \mathcal{I}^{\mu\gamma}/\mathcal{I}_1\mathcal{I}^{\mu\gamma}) \to \Gamma(A, \mathcal{I}^{\mu(\beta+\gamma)}/\mathcal{I}_1\mathcal{I}^{\mu(\beta+\gamma)})$ is surjective. The estimate on μ for $\beta = \gamma = 1$ also holds for all larger β and γ. As shown in the previous paragraph, for sufficiently large μ, $\Gamma(A, \mathcal{I}^{\mu\beta}/\mathcal{I}^{\mu\beta+\mu}) \to \Gamma(A, \mathcal{I}^{\mu\beta}/\mathcal{I}_1\mathcal{I}^{\mu\beta})$ is surjective, $\mu \geq 1$. Thus it suffices to show that

$$\Gamma(A, \mathcal{I}_1\mathcal{I}^{\mu\beta}/\mathcal{I}^{\mu\beta+\mu}) \otimes \Gamma(A, \mathcal{I}^{\mu\gamma}/\mathcal{I}^{\mu\gamma+\mu}) \to \Gamma(A, \mathcal{I}_1\mathcal{I}^{\mu(\beta+\gamma)}/\mathcal{I}^{\mu(\beta+\gamma)+\mu})$$

is surjective. Again choose μ large enough so that Lemma 7.3 implies that

$$\Gamma(A, \mathcal{I}_1\mathcal{I}^{\mu\beta}/\mathcal{I}_2\mathcal{I}^{\mu\beta}) \otimes \Gamma(A, \mathcal{I}^{\mu\gamma}/p_{i_2}\mathcal{I}^{\mu\gamma}) \to \Gamma(A, \mathcal{I}_1\mathcal{I}^{\mu(\beta+\gamma)}/\mathcal{I}_2\mathcal{I}^{\mu(\beta+\gamma)})$$

is surjective. Continuing in this fashion, we eventually reach the last sequence of (7.2) and thereby finish the proof of the lemma.∎

THEOREM 7.5. *For all sufficiently large $\mu \geq \mu_o$, μ_o depending only on the intersection matrix and the genera of the A_i, the canonical map*

$$\Gamma(A, (\mathcal{I}^{\mu})^{\beta}) \otimes_C \Gamma(A, (\mathcal{I}^{\mu})^{\gamma}) \to \Gamma(A, (\mathcal{I}^{\mu})^{\beta+\gamma})$$

is surjective for all β, $\gamma \geq 1$.

Proof: We first show that it suffices to prove that the image of $\Gamma(A,(\mathcal{I}^{\mu})^{\beta}) \otimes_{C} \Gamma(A,(\mathcal{I}^{\mu})^{\gamma})$ contains $\Gamma(A,(\mathcal{I}^{\mu})^{\delta})$ for some δ. $H^1(A,(\mathcal{I}^{\mu})^{a}) = 0$ for all sufficiently large μ and all $a \geq 1$ by Lemma 6.15. Thus the projection map $\Gamma(A,(\mathcal{I}^{\mu})^{a}) \to \Gamma(A,\mathcal{I}^{\mu a}/\mathcal{I}^{\mu a+\mu})$ is surjective. $\Gamma(A,\mathcal{I}^{\mu a}/\mathcal{I}^{\mu a+\mu}) \otimes \Gamma(A,\mathcal{I}^{\mu\gamma}/\mathcal{I}^{\mu\gamma+\mu}) \to \Gamma(A,\mathcal{I}^{\mu(a+\gamma)}/\mathcal{I}^{\mu(a+\gamma)+\mu})$ is surjective by Lemma 7.4. Thus

$\Gamma(A,(\mathcal{I}^{\mu})^{a}) \otimes_{C} \Gamma(A,(\mathcal{I}^{\mu})^{\gamma}) \to \Gamma(A,\mathcal{I}^{\mu(a+\gamma)}/\mathcal{I}^{\mu(a+\gamma)+\mu})$ is surjective. Then, knowing $\Gamma(A,(\mathcal{I}^{\mu})^{\delta})$ is in the image, we can let $a+\gamma = \delta - 1$ and conclude that $\Gamma(A,(\mathcal{I}^{\mu})^{\delta-1})$ is in the image.

Let $f_1,\ldots,f_s \in \Gamma(A,(\mathcal{I}^{\mu})^{\beta})$ generate $(\mathcal{I}^{\mu})^{\beta}$ as an $\mathcal{O}$-module in some neighborhood of A. We can verify the existence of such f_i as follows: $\Gamma(A,(\mathcal{I}^{\mu})^{\beta}) \to \Gamma(A,(\mathcal{I}^{\mu})^{\beta}/\mathcal{I}_1(\mathcal{I}^{\mu})^{\beta})$ is surjective for any choice of $\mathcal{I}_1$. Suppose then that $\mathcal{I}_1$ is the ideal sheaf of A_0. $\Gamma(A,(\mathcal{I}^{\mu})^{\beta}/\mathcal{I}_1(\mathcal{I}^{\mu})^{\beta})$ is the sheaf of germs of sections of a line bundle of Chern class $-A_0 \cdot (\mu\beta Z) \geq \mu\beta$ over A_0. Given a point $a \in A_0$, let $\tilde{f} \in \Gamma(A,(\mathcal{I}^{\mu})^{\beta}/\mathcal{I}_1(\mathcal{I}^{\mu})^{\beta})$ be non-zero near a as a section of the line bundle. $f \in \Gamma(A,(\mathcal{I}^{\mu})^{\beta})$ projecting onto $\tilde{f}$ will generate $(\mathcal{I}^{\mu})^{\beta}$ near a since it must vanish to the prescribed orders on the A_j near a and will have no other zeros near a. The $\mathcal{O}$-module map

$$\rho: \bigoplus_{s} (\mathcal{I}^{\mu})^{\gamma} \to (\mathcal{I}^{\mu})^{\beta+\gamma}$$

given by $(g_1,\ldots,g_s) \to \Sigma f_i g_i$ is then surjective. Let $\mathcal{K} = \ker \rho$.

$$0 \to \mathcal{K} \to \oplus (\mathcal{I}^{\mu})^{\gamma} \xrightarrow{\rho} (\mathcal{I}^{\mu})^{\beta+\gamma} \to 0$$

is exact. $\pi_*(\mathcal{K})$, the direct image of $\mathcal{K}$ under the resolving map π (which also exhibits A as exceptional), is coherent since

$$\pi_*(\mathcal{K}) = \ker \rho_*: \pi_*(\oplus(\mathcal{I}^{\mu})^{\gamma}) \to \pi_*((\mathcal{I}^{\mu})^{\beta+\gamma})$$

is the kernel of an $\mathcal{O}$-module map of coherent sheaves.

$$\begin{array}{ccccccccc} 0 & \to & \mathcal{K}(\mathcal{J}^{\mu})^{k} & \to & \oplus(\mathcal{J}^{\mu})^{\gamma+k} & \to & (\mathcal{J}^{\mu})^{\beta+\gamma+k} & \to & 0 \\ & & \downarrow \sigma & & \downarrow & & \downarrow \lambda & & \\ 0 & \to & \mathcal{K} & \to & \oplus(\mathcal{J}^{\mu})^{\gamma} & \to & (\mathcal{J}^{\mu})^{\beta+\gamma} & \to & 0, \end{array}$$

with the vertical maps the inclusion maps, is commutative. The verification that the first line is exact is the same as the verification that (5.5) was exact.

$$\begin{array}{ccccc} \Gamma(A, \oplus(\mathcal{J}^{\mu})^{\gamma+k}) & \to & \Gamma(A,(\mathcal{J}^{\mu})^{\beta+\gamma+k}) & \to & H^1(A,\mathcal{K}(\mathcal{J}^{\mu})^{k}) \\ \downarrow & & \downarrow \lambda_* & & \downarrow \sigma_* \\ \Gamma(A, \oplus(\mathcal{J}^{\mu})^{\gamma}) & \overset{\rho_*}{\to} & \Gamma(A,(\mathcal{J}^{\mu})^{\beta+\gamma}) & \to & H^1(A,\mathcal{K}) \end{array}$$

is commutative with exact rows. By Theorem 5.4, σ_* is the zero map for sufficiently large k. Then given $h \in \Gamma(A,(\mathcal{J}^{\mu})^{\beta+\gamma+k})$, $\lambda_*(h) = \rho_*(g)$ for some g, by exactness. Letting $\delta = \beta+\gamma+k$, we have that the image of $\Gamma(A,(\mathcal{J}^{\mu})^{\beta}) \otimes_C \Gamma(A,(\mathcal{J}^{\mu})^{\gamma})$ contains $\Gamma(A,(\mathcal{J}^{\mu})^{\delta})$ as required. ▮

Recall that $\mathcal{J} = \Pi\, p_i^{r_i}$ with $p_i = \mathcal{I}d(A_i)$. Let $\theta = \max(r_i)$. The following result is immediate from Theorem 7.5.

COROLLARY 7.6. *Let* R *be the ring* $\Gamma(A,\mathcal{O}) = \mathcal{O}_p$, *the germs of holomorphic functions at the singularity* p. *Let* m *be the maximal ideal of* R. *Then for all* $\nu \geq 1$, $m^{\nu\mu\theta} \subset \Gamma(A,(\mathcal{J}^{\mu})^{\nu}) \subset m^{\nu}$.

LEMMA 7.7. *Let* $p \in V$, *an analytic subvariety. Let* m *be the maximal ideal in the ring of germs of holomorphic functions at* p. *Given* $\nu \geq 1$, *let* $f_1, \ldots, f_s$ *be germs of functions in* m^{ν} *whose images in* $m^{\nu}/m^{2\nu}$ *span* $m^{\nu}/m^{2\nu}$ *as a vector space. Then any* $g \in m^{\nu}$ *can be written as a convergent power series with the* f_i *as variables.*

Proof: First consider the case where p is a regular point. Suppose that $z_1, \ldots, z_t$ are local coordinates near p and that the f_i are in fact $\{z^\alpha \mid \nu \leq |\alpha| \leq 2\nu - 1\}$, α a multi-index. Suppose $m^\nu \ni g = \Sigma\, a_\beta z^\beta$, $|\beta| \geq \nu$. $\Sigma\, a_\beta z^\beta$ is a convergent power series so say it converges for some $\zeta = (\zeta_1, \ldots, \zeta_t)$ with no $\zeta_i = 0$. $\{a_\beta \zeta^\beta\}$ is then a bounded set. $G(z) = g(\zeta z) = g(\zeta_1 z_1, \ldots, \zeta_t z_t) = \Sigma\, (a_\beta \zeta^\beta) z^\beta$ has bounded coefficients. It suffices to represent G as a power series in the z^α, for then we may make the inverse change of variables and recover g. But $z^\beta = (\Pi\, z^\alpha)_\beta$, for appropriate (although not unique) α. Hence $G(z) = \Sigma\, (a_\beta \zeta^\beta)(\Pi\, z^\alpha)_\beta$, with bounded coefficients. Thus the power series is convergent.

Still suppose that p is a regular point, but now consider arbitrary $f_1, \ldots, f_s$. Choosing a basis among the images of the f_i in $m^\nu/m^{2\nu}$ and making a linear change among the f_i, we may write the f_i as $f_\alpha = z^\alpha + \Sigma\, a_{\alpha\beta} z^\beta$, $|\beta| \geq 2\nu$. From the previous paragraph, $f_\alpha = z^\alpha + h_\alpha(z^{\alpha_1}, \ldots, z^{\alpha_s})$, with $h_\alpha(\zeta_1, \ldots, \zeta_s)$ a convergent power series with no constant or linear terms. By the inverse function theorem, $\{f_\alpha(\zeta) = \zeta_\alpha + h_\alpha(\zeta_1, \ldots, \zeta_s)\}$ may be inverted to give $\zeta_\alpha = H_\alpha(f_1, \ldots, f_s)$ where H_α is holomorphic. Letting $\zeta_\alpha = z^\alpha$ gives z^α as a holomorphic function of the f_α. Convergent power series in the f_α yield m^ν because convergent power series in the z^α yield m^ν.

Finally consider the general case where p may be singular. Let $z_1, \ldots, z_t$ be ambient coordinate functions. We are given $f_1, \ldots, f_s$ whose images span $m^\nu/m^{2\nu}$. Then there are linear functions L_α such that on V, $z^\alpha \equiv L_\alpha(f) \bmod m^{2\nu}$, $\nu \leq \alpha \leq 2\nu - 1$. Let M be the ambient maximal ideal. The projection map $M^\nu \to m^\nu$ is surjective for all ν.

Thus there are ambient functions $H_\alpha(z) \in M^{2\nu}$ such that $L_\alpha(f) \equiv z^\alpha + H_\alpha(z) \bmod (\mathcal{I}d\, V)_p$. Then the $z^\alpha + H_\alpha(z)$ form a basis of

$M^\nu/M^{2\nu}$. Hence by the first part of this proof, power series in the $z^\alpha + H_\alpha(z)$ yield M^ν. Thus power series in the f_i yield m^ν.∎

LEMMA 7.8. *Let $\mu \geq \mu_0$, μ_0 to depend only on the intersection matrix and genera of the* A_i. *Let* $f_1, \ldots, f_s \in \Gamma(A, \mathcal{J}^\mu)$ *have images in* $\Gamma(A, \mathcal{J}^\mu/\mathcal{J}^{2\mu})$ *which form a basis. Then there is a neighborhood* U *of* A *such that* $F = (f_1, \ldots, f_s): U \to W$ *is a proper map onto a subvariety* W *of a polydisc in* C^s. F *is biholomorphic off* A. $\Gamma(A, \mathcal{J}^\mu)$ *is isomorphic to the maximal ideal in* ${}_W\mathcal{O}_0$.

Proof: First require that μ be large enough for Lemma 6.18 to hold. Then, as homogeneous coordinates, $(f_1, \ldots, f_s)$ embeds U–A in P^{s-1} for some U. Thus F is biholomorphic on U–A.

We see that F is proper as follows. Let $\pi: U \to V$ exhibit A as exceptional, $\pi(A) = p$. Let m be the maximal ideal of ${}_V\mathcal{O}_p$. By Corollary 7.6, $\Gamma(A, \mathcal{J}^\mu) \supset m^{\mu\theta}$ and $\Gamma(A, (\mathcal{J}^\mu)^{2\mu\theta}) \subset m^{2\mu\theta}$. Polynomials in the f_i project onto $\Gamma(A, \mathcal{J}^\mu/(\mathcal{J}^\mu)^{2\mu\theta})$ by Lemma 7.4. Therefore there are polynomials in the f_i whose images in $m^{\mu\theta}/m^{2\mu\theta}$ form a basis. Let $z_1, \ldots, z_t$ be ambient local coordinates for V near p. By Lemma 7.7, $z_j^{\mu\theta}$, $1 \leq j \leq t$, can be written as a convergent power series in the f_i, $z_j^{\mu\theta} = \psi_j(f_1, \ldots, f_s)$. Let R be a small enough neighborhood of the origin in C^s so that the ψ_i coverage in R. $\psi = (\psi_1, \ldots, \psi_t): R \to C^t$ is then a holomorphic map. $\psi \circ F: F^{-1}(R) \to C^t$ is holomorphic and given by $(z_1^{\mu\theta}, \ldots, z_t^{\mu\theta})$. We can choose a polydisc Δ in C^t so that $\pi = (z_1, \ldots, z_t)$ from above is proper on $\pi^{-1}(\Delta)$. Since $\lambda: z_i \to z_i^{\mu\theta}$ is a proper map, for any polydisc Δ' such that $\lambda(\Delta) \supset \Delta'$ and $(\lambda \circ \pi)^{-1}(\Delta') \subset F^{-1}(R)$, $\psi \circ F = \lambda \circ \pi: (\lambda \circ \pi)^{-1}(\Delta') \to \Delta'$ is a proper map. F is then a proper map from $(\lambda \circ \pi)^{-1}(\Delta')$, a neighborhood of A, to $\psi^{-1}(\Delta')$, a neighborhood of the origin, as needed. By the proper mapping theorem, Theorem V.C.5 of G & R, the image W is a subvariety.

Finally, since polynomials in the f_i span $m^{\mu\theta}/m^{2\mu\theta}$, by Lemma 7.7, convergent power series in the f_i yield $m^{\mu\theta}$. Since polynomials in the f_i project onto $\Gamma(A, \mathcal{I}^{\mu}/(\mathcal{I}^{\mu})^{\mu\theta})$ and $\Gamma(A, (\mathcal{I}^{\mu})^{\mu\theta}) \subset m^{\mu\theta}$, convergent power series in the f_i yield $\Gamma(A, \mathcal{I}^{\mu})$. But ${}_W\mathcal{O}_0$ is nothing more than the restriction to W of convergent ambient power series, i.e. ${}_W\mathcal{O}_0$ is exactly those functions in a neighborhood of 0 which can be written as convergent power series in the f_i plus a constant term. ∎

W is, of course, not normal in general. 0 is an isolated singularity and $F: U \to W$ is a resolution. U can be obtained by a single ambient quadratic transformation of $\mathbf{C}^s$, as defined below. Compare this with our highly non-canonical resolution of Chapter II.

DEFINITION 7.1. *Let $(z_1, \ldots, z_s)$ be coordinates for $\mathbf{C}^s$. A quadratic transformation $\pi: S \to \mathbf{C}^s$ at the origin is given as follows. S is a line bundle over $\mathbf{P}^{s-1}$ and π maps the fibres of S onto the corresponding lines through the origin in $\mathbf{C}^s$. In terms of coordinates, if $u_1, \ldots, u_{s-1}$ are inhomogeneous coordinates in $\mathbf{P}^{s-1}$ with $(u_1, \ldots, u_{s-1}, 1)$ the corresponding homogeneous coordinates and v is the fibre coordinate in S, $\pi(u_1, \ldots, u_{s-1}, v) = (u_1 v, u_2 v, \ldots, u_{s-1} v, v)$. If $(u_1', \ldots, 1, \ldots, u_s')$ are different inhomogeneous coordinates and v' is the corresponding fibre coordinate, $\pi(u_1', \ldots, u_s', v') = (u_1' v', \ldots, v', \ldots, u_s' v')$. $v' = u_k v$ for some k. $v = u_s' v'$ and $u_k = 1/u_s'$.*

If D is an open subset of $\mathbf{C}^s$ with $0 \in D$, $\pi: \pi^{-1}(D) \to D$ is the quadratic transformation of D at the origin. Let D' denote $\pi^{-1}(D)$.

Suppose that W is a subvariety of D such that $0 \in W$ but 0 is not an isolated point of W. Let W' be the closure of $\pi^{-1}(W-0)$. $\pi^{-1}(W)$ is a subvariety of D' with $\pi^{-1}(0)$ as a proper subvariety. Thus W' is the closure of those components of the regular points of $\pi^{-1}(W)$ which contain the regular points of $\pi^{-1}(W-0)$. Thus W' is a subvariety of D'.

What points of $\pi^{-1}(0)$ are in W'?
$\pi^{-1}(z_1, \ldots, z_s) = (z_1/z_s, \ldots, z_{s-1}/z_s, z_s)$ for $z_s \neq 0$. Thus for $z \neq 0$, $\pi^{-1}(z)$ lies in the fibre over the point $\mathbf{P}^{s-1}$ having $(z_1, \ldots, z_s)$ as homogeneous coordinates. Hence the points in $\pi^{-1}(0) \cap W'$ are the

points in P^{s-1} which are limit points in homogeneous coordinates of $(z_1, \ldots, z_s)$ as $z \to 0$ in W.

Now consider the W of Lemma 7.8 and an ambient quadratic transformation at $0 \in W \subset C^s$. A point in $W' \cap \pi^{-1}(0)$ must correspond to a limit point in P^{s-1} of $(f_1(x), \ldots, f_s(x))$ as $x \to A$. But, by Lemma 6.18, $(f_1, \ldots, f_s)$ embeds a neighborhood of A in P^{s-1}. Thus $W' \cap \pi^{-1}(0) \approx A$. In fact we have that U and W′ are biholomorphic for consider a point $a \in A$. Assume, without loss of generality, that f_s generates $\mathcal{I}^\mu$ near a. Then $(f_1/f_s, \ldots, f_{s-1}/f_s, f_s)$ is a biholomorphic map from a neighborhood of a to a neighborhood of the corresponding point in W′, written in inhomogeneous coordinates with f_s the fibre coordinate. Thus $\pi|_{W'}: W' \to W$ is a resolution of W with $W' \approx U$ and $\pi^{-1}(0) \cap W' \approx A$. We shall identify U with W′ and A with $\pi^{-1}(0) \cap W'$.

We now wish to get defining equations for W′ in D′, using the C-algebra structure of ${}_W\mathcal{O}_o$. In our proof of Theorem 7.1, we will require estimates on the defining equations in terms of m^ν. By Corollary 7.6, it will suffice to get estimates in terms of $\Gamma(A, \mathcal{I}^{\mu})^{\nu})$. Suppose then that $\zeta_1, \ldots, \zeta_s$ are ambient local coordinates near 0 in W. $\zeta_i = f_i$. Suppose also that $g(\zeta) = \Sigma\, a_\alpha \zeta^\alpha \in (\mathcal{I}d\,(W))_o$ or equivalently that $g(\zeta) = 0$ for $\zeta \in W$. Locally,

$$\pi^* g = g(u_1 v, \ldots, u_{s-1} v, v) = \Sigma\, a_\alpha v^{|\alpha|} u^{\alpha'}$$

where α' is the multi-index obtained from α by deleting the last entry in α. If $a_\alpha = 0$ for $|\alpha| < \nu$, $\pi^* g/v^\nu$ is holomorphic in a neighborhood of $\pi^{-1}(0)$. $\pi^* g/v^\nu$ vanishes on $\pi^{-1}(W-0)$ and hence on W′. $\pi^{-1}(0)$ is given locally by $\{v = 0\}$. Thus $\text{loc}\ \pi^* g/v^\nu \cap \pi^{-1}(0)$ depends only on the lowest order homogeneous terms of the power series expansion of g. Moreover, $\text{loc}\ \pi^* g/v^\nu \cap \pi^{-1}(0)$ is given in homogeneous coordinates by the lowest order terms in the power series expansion for g. Let M be the ambient maximal ideal. We have shown that if

$g, g' \in M^{\nu} - M^{\nu+1}$ and $g - g' \in M^{\nu+1}$, $\overset{*}{\pi} g$ and $\overset{*}{\pi} g'$ yield the same equation for A.

We now determine how large a ν is needed in order for these equations for A to have exactly A as their common zeros. Let $\mathcal{J}$ be the ideal sheaf of A. Recall that $\theta = \max(r_i)$. If $f_i \in \Gamma(A, \mathcal{J}\mathcal{I}^{\mu}), f_i^{\mu\theta} \in \Gamma(A, \mathcal{I}^{\mu(\mu\theta+1)})$. Hence by Theorem 7.5, $f_i^{\mu\theta} = \Sigma\, b_k(\Pi\, h_k)$ where we have a finite summation, $h_k \in \Gamma(A, \mathcal{I}^{\mu})$ and each product has precisely $(\mu\theta+1)$ factors. By Lemma 7.8, we may express the h_k as power series s_k in the f_j. Hence $g(\zeta) = \zeta_i^{\mu\theta} - \Sigma\, b_k(\Pi s_k(\zeta))$ vanishes when $\zeta_j = f_j$, all j, i.e. on W. Then, in homogeneous coordinates, $\zeta_i^{\mu\theta} = 0$ is an equation for A. $\zeta_i = 0$ is then also an equation for A. We still have left to consider those $f_i \notin \Gamma(A, \mathcal{J}\mathcal{I}^{\mu})$. For sufficiently large μ, $H^1(A, \mathcal{J}\mathcal{I}^{\mu}) = 0$ so we may assume that $f_1, \ldots, f_t$ project onto a basis for $H^1(A, \mathcal{I}^{\mu}/\mathcal{J}\mathcal{I}^{\mu})$ and $f_{t+1}, \ldots, f_s \in \Gamma(A, \mathcal{J}\mathcal{I}^{\mu})$. As shown in the proof of Lemma 6.18, $f_1, \ldots, f_t$ embed A in P^{t-1} and this embedding depends only on the images of $f_1, \ldots, f_t$ in $\Gamma(A, \mathcal{I}^{\mu}/\mathcal{J}\mathcal{I}^{\mu})$. We shall later prove the following lemma, which is very similar to Lemma 7.3.

LEMMA 7.9. *Let* A *be as in Lemma 7.2. Let* L *and* M *be line bundles over* A *with* $c_i(L) \geq 4g_i + 2\omega + 2$ *and* $c_i(M) \geq c_i(L) + (2g_i - 2) + \omega + 2$. *Let* S *be a codimension* 1 *subspace of* $\Gamma(A, \mathcal{L})$ *such that given* $x \in A$, *there exists an* $f \in S$ *such that* $f(x) \neq 0$. *Then the canonical map* $S \otimes_C \Gamma(A, \mathcal{M}) \to \Gamma(A, \mathcal{L} \otimes \mathcal{M})$ *is surjective.*

Regard $\Gamma(A, \mathcal{I}^{\mu}/\mathcal{J}\mathcal{I}^{\mu})$ as sections of a line bundle L over A. Suppose $w \in P^{t-1}$ and w does not lie in the image of $(f_1, \ldots, f_t)$. Using homogeneous coordinates and making a linear change of coordinates, we may assume that $w = (1, 0, \ldots, 0)$. w does not lie in the image of A means that there does not exist a point $x \in A$ such that $f_1(x) \neq 0$, $f_2(x) = 0, \ldots, f_t(x) = 0$, as sections of L. Let S be the subspace of $\Gamma(A, \mathcal{I}^{\mu}/\mathcal{J}\mathcal{I}^{\mu}) = \Gamma(A, \mathcal{L})$ spanned by $f_2, \ldots, f_t$. Choose μ sufficiently large so that $c_i(\mathcal{I}^{\mu}/\mathcal{J}\mathcal{I}^{\mu}) \geq 4g_i + 2\omega + 2$. Let $M = L \otimes L$. Then

$\Gamma(A,\mathfrak{L}) \otimes \Gamma(A,\mathfrak{L}) \to \Gamma(A,\mathfrak{M})$ is surjective by Lemma 7.3.
$S \times \Gamma(A,\mathfrak{M}) \to \Gamma(A,\mathfrak{L}\otimes\mathfrak{L}\otimes\mathfrak{L})$ is surjective by Lemma 7.9. In particular, f_1^3 lies in the image, i.e., $f_1^3 - \Sigma\, a_{ijk} f_i f_j f_k \in \Gamma(A, \mathcal{J}\mathcal{J}^{3\mu})$ for appropriate constants a_{ijk} with $i \neq 1$, $i,j,k \leq t$. As shown in the proof of Lemma 7.4, for sufficiently large μ,

$$\Gamma(A,\mathcal{J}\mathcal{J}^{\mu}/\mathcal{J}^{2\mu}) \otimes \Gamma(A,\mathcal{J}^{\mu}/\mathcal{J}^{2\mu}) \to \Gamma(A,\mathcal{J}\mathcal{J}^{2\mu}/\mathcal{J}^{3\mu})$$

and

$$\Gamma(A,\mathcal{J}\mathcal{J}^{2\mu}/\mathcal{J}^{3\mu}) \otimes \Gamma(A,\mathcal{J}^{\mu}/\mathcal{J}^{2\mu}) \to \Gamma(A,\mathcal{J}\mathcal{J}^{3\mu}/\mathcal{J}^{4\mu})$$

are surjective.

Thus $f_1^3 - \Sigma\, a_{ijk} f_i f_j f_k - \Sigma\, b_{\alpha\beta\gamma} f_\alpha f_\beta f_\gamma \in \Gamma(A,\mathcal{J}^{4\mu})$ with $t+1 \leq \alpha \leq s$, $1 \leq \beta, \gamma \leq s$. Then for appropriate $n \in M^4$, $g(\zeta) = \zeta_1^3 - \Sigma\, a_{ijk}\zeta_i\zeta_j\zeta_k - \Sigma\, b_{\alpha\beta\gamma}\zeta_\alpha\zeta_\beta\zeta_\gamma - n(\zeta)$ vanishes on W and gives $\zeta_1^3 - \Sigma\, a_{ijk}\zeta_i\zeta_j\zeta_k - \Sigma\, b_{\alpha\beta\gamma}\zeta_\alpha\zeta_\beta\zeta_\gamma$ as an equation for A. We know that $\zeta_\alpha = 0$ on A since $\alpha \geq t+1$. Also, $(1,0,\ldots,0)$ does not satisfy $\zeta_1^3 - \Sigma\, a_{ijk}\zeta_i\zeta_j\zeta_k$ since $i \neq 1$. Thus from the C-algebra structure of $\Gamma(A,\mathcal{J}^{\mu})/\Gamma(A,\mathcal{J}^{4\mu})$ and from $\zeta_\alpha = 0$ for $\alpha \geq t+1$, which we derived from $\Gamma(A,\mathcal{J}^{\mu})/\Gamma(A,(\mathcal{J}^{\mu})^{\mu\theta+1})$, we have found a complete set of defining equations for A. (We do not claim that these functions generate $\mathcal{J}d(A)$.)

Proof of Lemma 7.9. Since elements of S have no common zeroes, there is an $f \in S$ such that f does not vanish at any singular point and f does not vanish identically on any A_i. Let $\{P_{ij}\}$ be the points on A_i where f vanishes and let a_{ij} be the order of the zero of f at P_{ij}. As in the proof of Lemma 7.3, $f \otimes \Gamma(A,\mathfrak{M})$ has as image in $\Gamma(A,\mathfrak{L}\otimes\mathfrak{M})$ the subspace Q whose elements vanish to order at least a_{ij} at the P_{ij}. Since $\sum_j a_{ij} = c_i(L)$, there are at most $c_i(L)$ of the P_{ij} for any given i. Since $c_i(M) \geq c_i(L) + (2g_i - 2) + \omega + 1$, by Lemma 7.2 we can find $g \in \Gamma(A,\mathfrak{M})$ with any specified values at the P_{ij}. There exists an $h \in S$

such that $h(P_{ij}) \neq 0$ for all P_{ij}. Thus with the appropriate h, $h \otimes g$ will take on any specified values at the P_{ij}. Hence, knowing that Q is in the image, it suffices to find elements of the image which vanish at the P_{ij} to exactly given orders b_{ij}, $1 \leq b_{ij} \leq a_{ij}$. Given such a set of b_{ij}, choose c_{ij} so that $1 \leq c_{ij} \leq b_{ij}$, $c_{ij} = 1$ if $\sum_j b_{ij} \leq c_i(L) - (2g_i - 2) -1 - \omega - 2$ and $\sum_j c_{ij} \geq (2g_i - 2) + 1 + \omega + 2$ if $\sum_j b_{ij} > c_i(L) - (2g_i-2) -1-\omega-2$. Let $d_{ij} = b_{ij} - c_{ij}$. We can find as follows an $h \in S$ such that at P_{ij}, h vanishes to order exactly d_{ij} or $d_{ij}+1$. $\sum_j d_{ij} \leq c_i(L) - (2g_i-2) - 1 - \omega - 2$. Therefore by Lemma 7.2, the space S_1 of sections of L vanishing to order at least $d_{ij}+2$ at a given P_{ij} and at least d_{ij} at the other P_{ij} has codimension 2 in the space S_2 of sections of L vanishing to order at least d_{ij} at all the P_{ij}. S is of codimension 1 in $\Gamma(A,\mathcal{L})$. Therefore $S \cap S_1$ is of codimension at least 1 in $S \cap S_2$. There are only a finite number of P_{ij} so that $\{S \cap S_1\}$ cannot exhaust $S \cap S_2$. Thus $h \in S \supset (S \cap S_2)$ exists. $\sum_j c_{ij} \leq c_i(L)$ so there is a $g \in \Gamma(A,\mathfrak{M})$ so that g vanishes to order exactly c_{ij} or $c_{ij}-1$ (depending on whether h vanishes to order d_{ij} or $d_{ij}+1$) at P_{ij}. $h \otimes g$ vanishes to order exactly b_{ij}, as desired.∎

We shall now derive generators near A for $\mathfrak{Id}(W')$ in D'. First consider a regular point q in say A_1. Choose local coordinates (x,y) on U so that $q = (0,0)$ and $A_1 = \{y=0\}$. Let $d = \mu r_1$ so that the f_j vanish to d^{th} order on A_1. For sufficiently large μ, $H^1(A,\mathcal{J}\mathfrak{I}^\mu) = 0$, so $\Gamma(A,\mathfrak{I}^\mu) \to \Gamma(A,\mathfrak{I}^\mu/\mathcal{J}\mathfrak{I}^\mu)$ is surjective. $\Gamma(A,\mathfrak{I}^\mu/\mathfrak{I}^{2\mu}) \approx \Gamma(A,\mathfrak{I}^\mu)/\Gamma(A,\mathfrak{I}^{2\mu})$. So, making a non-singular linear change among the f_j, we may assume that $f_1,\ldots,f_{a-1}$ project onto a basis for $\Gamma(A,\mathfrak{I}^\mu/\mathcal{J}\mathfrak{I}^\mu)$ and that $f_a,\ldots,f_s \in \Gamma(A,\mathcal{J}\mathfrak{I}^\mu)$. $\Gamma(A,\mathfrak{I}^\mu/\mathcal{J}\mathfrak{I}^\mu)$ corresponds to sections of a line bundle with $c_i(\mathfrak{I}^\mu/\mathcal{J}\mathfrak{I}^\mu) \to \infty$ as $\mu \to \infty$. Then by our usual patching construction, $\Gamma(A,\mathfrak{I}^\mu/\mathcal{J}\mathfrak{I}^\mu) \to \Gamma(A,\mathfrak{I}^\mu/p_1\mathfrak{I}^\mu)$ is surjective for sufficiently large μ. Thus we can choose $f_1,\ldots,f_b$ to project onto a basis of $\Gamma(A,\mathfrak{I}^\mu/p_1\mathfrak{I}^\mu)$ and moreover such that near $(0,0)$

(7.3) $$f_1 = y^d[1 + g_1(x)] + y^{d+1}h_1(x,y) \quad ,$$

g_1 and h_1 holomorphic with $g_1(0) = 0$. $f_1 \epsilon \Gamma(A, p_2 \cdots p_n \mathcal{I}^\mu)$.

(7.4) $$f_2 = y^d[x + g_2(x)] + y^{d+1}h_2(x,y)$$

with g_2 and h_2 holomorphic, $g_2(0) = g_2'(0) = 0$.

As elements of $\Gamma(A, \mathcal{I}^\mu / p_1 \mathcal{I}^\mu)$, $f_3, \ldots, f_b$ should all have at least a second-order zero at q. $f_{b+1}, \ldots, f_{a-1} \epsilon \Gamma(A, p_1 \mathcal{I}^\mu)$.

$\Gamma(A, \mathcal{J}\mathcal{I}^\mu / \mathcal{J}^2 \mathcal{I}^\mu)$ represents sections of a line bundle, so for sufficiently large μ, we may take

(7.5) $$f_a = y^{d+1}[1 + g_a(x)] + y^{d+2}h_a(x,y), \quad g_a(0) = 0 .$$

The ambient quadratic transformation is given locally by $\pi(u_2, \ldots, u_s, v) = (v, u_2 v, \ldots, u_s v) = (\zeta_1, \ldots, \zeta_s)$. $\zeta_i = f_i(x,y)$. $u_2 = f_2/f_1 = x + \ldots$ and $u_a = f_a/f_1 = y + \ldots$ are local coordinates for U, which is a submanifold of D′, the ambient space. We must express the coordinates for D′ in terms of linear combinations of u_1 and u_2 and higher order terms in all of the variables. Recall that $\theta = \max(r_i)$.

$f_1^{\theta\mu}[f_1^{d+1} - f_a^d] \epsilon \Gamma(A, (\mathcal{I}^\mu)^{\theta\mu + d + 1})$ since $f_a^d \epsilon \Gamma(A, \mathcal{J}^d \mathcal{I}^{d\mu})$, $f_1^\theta \epsilon \Gamma(A, p_2^\theta \cdots p_n^\theta \mathcal{I}^{\theta\mu})$ and $\mathcal{J}^d p_2^{\theta\mu} \cdots p_n^{\theta\mu} \subset \mathcal{I}^\mu$. By Lemma 7.8 and Theorem 7.5, we can write $f_1^{\theta\mu}[f_1^{d+1} - f_a^d]$ as a convergent power series in the f_j with no terms of order less than $\theta\mu + d + 1$. But $f_1^{\theta\mu}[f_1^{d+1} - f_a^d] = y^{d[\theta\mu + d + 1]} h(x,y)$ with $h(0,0) = 0$ by (7.3) and (7.5). Thus the non-zero homogeneous terms of order $\theta\mu + d + 1$ in the power series expansion for $f_1^{\theta\mu}[f_1^{d+1} - f_a^d]$ cannot include

$f_1^{\theta\mu+d+1} = y^{d[\theta\mu+d+1]}\,[1+\ldots]$ since $f_j = y^d k_j(x,y)$, $k_j(0,0) = 0$ for $j > 1$. Then

$$\zeta_1^{\theta\mu+d+1} - \zeta_1^{\theta\mu}\,\zeta_a^d - \Sigma\, a_\alpha\, \zeta^\alpha \in \mathcal{I}(W) \tag{7.6}$$

with $|\alpha| \geq \theta\mu+d+1$ and $a_\alpha = 0$ for $\alpha = (\theta\mu+d+1, 0, \ldots, 0)$.

Under the quadratic transformation, (7.6) yields the following equation for W'.

$$v^{\theta\mu+d+1} - v^{\theta\mu+d}\,u_a^d - \Sigma\, a_\alpha v^{|\alpha|}\, u^{\alpha'} = 0\,, \quad |\alpha| \geq \theta\mu+d+1\,,$$

where α' is the multi-index obtained from α by deleting the first entry. Thus $|\alpha| = \theta\mu+d+1$, $\alpha' = (0, \ldots, 0)$ does not occur in the summation. Dividing out the $v^{\theta\mu+d}$ we get

$$v - u_a^d - \Sigma\, a_\alpha v^{|\alpha| - \theta\mu - d} u^{\alpha'} = 0 \tag{7.7}$$

and the summation contains only terms of at least second order. Thus v depends on quadratic and higher order terms.

Now consider $f_3, \ldots, f_{a-1}$, whose images have second order zeroes at q in $\Gamma(A, \mathcal{I}^\mu/p_1\mathcal{I}^\mu)$. Let $\mathcal{L} = \mathcal{M}$ be the subsheaf of $\mathcal{I}^\mu/\mathcal{J}\mathcal{I}^\mu$ of elements which vanish at q as sections of the corresponding bundle. By Lemma 7.3, given j, $3 \leq j \leq a-1$, there are constants a_{ke}, $2 \leq k,e \leq a-1$ such that $f_1 f_j - \Sigma\, a_{ke} f_k f_e$ projects onto the zero section in $\mathcal{L} \otimes \mathcal{M}$. Thus $f_1 f_j - \Sigma\, a_{ke} f_k f_e \in \Gamma(A, \mathcal{J}\mathcal{I}^{2\mu})$. Look at the image of $f_1 f_j - \Sigma\, a_{ke} f_k f_e$ in $\Gamma(A, \mathcal{J}\mathcal{I}^{2\mu}/\mathcal{J}^2\mathcal{I}^{2\mu})$. For appropriate constants α and β, $f_1 f_j - \Sigma\, a_{ke} f_k f_e - \alpha f_1 f_a - \beta f_2 f_a$ will have a second order zero as a section of the line bundle corresponding to $\mathcal{J}\mathcal{I}^{2\mu}/\mathcal{J}^2\mathcal{I}^{2\mu}$. For sufficiently large μ, $H^1(A, \mathcal{J}^2\mathcal{I}^\mu) = 0$. $\Gamma(A, \mathcal{J}\mathcal{I}^\mu) \to \Gamma(A, \mathcal{J}\mathcal{I}^\mu/\mathcal{J}^2\mathcal{I}^\mu)$ is then surjective so $\Gamma(A, \mathcal{J}\mathcal{I}^{\mu}/\mathcal{I}^{2\mu}) \to \Gamma(A, \mathcal{J}\mathcal{I}^{\mu}/\mathcal{J}^{2}\mathcal{I}^{\mu})$ is surjective. Then by Lemma 7.3, there are constants $b_{\nu\tau}$, $2 \leq \nu \leq a-1$, $a \leq \tau$ such that

$f_1 f_j - \Sigma a_{ke} f_k f_e - \alpha f_1 f_a - \beta f_2 f_a - \Sigma b_{\nu\tau} f_\nu f_\tau \in \Gamma(A, \mathcal{J}^2 \mathcal{I}^{2\mu})$. The proof of Lemma 7.4 is easily modified to show that for sufficiently large μ,

$\Gamma(A, \mathcal{J}\mathcal{I}^{\mu}/\mathcal{I}^{2\mu}) \otimes \Gamma(A, \mathcal{J}\mathcal{I}^{\mu}/\mathcal{I}^{2\mu}) \to \Gamma(A, \mathcal{J}^2 \mathcal{I}^{2\mu}/\mathcal{J}\mathcal{I}^{3\mu})$ is surjective. So finally there exist constants c_{mn}, $a \leq m,n$ such that

$f_1 f_j - \Sigma a_{ke} f_k f_e - \alpha f_1 f_a - \beta f_2 f_a - \Sigma b_{\nu\tau} f_\nu f_\tau - \Sigma c_{mn} f_m f_n \in \Gamma(A, \mathcal{I}^{3\mu})$,

with $k, e, \nu, \tau, m, n \geq 2$. Then for an appropriate $g(\zeta)$, having only third and higher order terms in its power series expansion

$$\zeta_1 \zeta_j - \Sigma a_{ke} \zeta_k \zeta_e - \alpha \zeta_1 \zeta_a - \beta \zeta_2 \zeta_a - \Sigma b_{\nu\tau} \zeta_\nu \zeta_\tau - \Sigma c_{mn} \zeta_m \zeta_n - g(\zeta) \in \mathcal{I}d(W).$$

This yields the following equation for W'.

$$u_j - \Sigma a_{ke} u_k u_e - \alpha u_a - \beta u_2 u_a - \Sigma b_{\nu\tau} u_\nu u_\tau$$

(7.8)

$$- \Sigma c_{mn} u_m u_n - \frac{1}{v^2} g(v_1 u_2 v, \ldots, u_s v) = 0$$

The power series expansion for $\frac{1}{v^2} g(v, u_2 v, \ldots, u_s v)$ contains v in every monomial. v depends on quadratic terms by (7.7). Hence (7.8) expresses u_j as αu_a + (quadratic terms), as desired.

For $j \geq a$, $f_j \in \Gamma(A, \mathcal{J}\mathcal{I}^{\mu})$. $f_1 f_j \in \Gamma(A, \mathcal{J}\mathcal{I}^{2\mu})$. Thus we may repeat the argument just given above for $f_1 f_j - \Sigma a_{ke} f_k f_e$ and obtain an equation like (7.8) for u_j.

We now must find generators for $\mathcal{I}d(W')$ in D' near a singular point q. Say $q = A_1 \cap A_2$. Choose local coordinates (x,y) near $q = (0,0)$ so that $A_1 = \{y = 0\}$, $A_2 = \{x = 0\}$. Let $d = \mu r_1$ and $e = \mu r_2$ so $\mathcal{I}^{\mu}$ is generated near q by $y^d x^e$. Making a non-singular linear change among the f_j we may choose, by first choosing the appropriate image in $\Gamma(A, \mathcal{I}^{\mu}/\mathcal{J}\mathcal{I}^{\mu})$,

(7.9) $$f_1 = y^d x^e [1 + g_1(x) + h_1(y)] + y^{d+1} x^{e+1} k_1(x,y)$$

with g_1, h_1, and k_1 holomorphic, $g_1(0) = h_1(0) = 0$. We also require that $f_1 \in \Gamma(A, p_3 \ldots p_n \mathcal{I}^\mu)$.

(7.10) $$f_2 = y^d x^e [x + g_2(x)] + y^{d+1} x^{e+1} k_2(x,y)$$

with g_2 and k_2 holomorphic, $g_2(0) = g_2'(0) = 0$ and $f_2 \in \Gamma(A, p_2 p_3 \ldots p_n \mathcal{I}^\mu)$.

(7.11) $$f_3 = y^d x^e [y + h_3(y)] + y^{d+1} x^{e+1} k_3(x,y)$$

with h_3 and k_3 holomorphic, $h_3(0) = h_3'(0) = 0$ and $f_3 \in \Gamma(A, p_1 p_3 p_4 \ldots p_n \mathcal{I}^\mu)$.

The images of $f_4, \ldots, f_s$ in $\Gamma(A, \mathcal{I}^\mu / p_1 \mathcal{I}^\mu)$ should vanish at q to at least second order. $u_2 = f_2/f_1$ and $u_3 = f_3/f_1$ are, of course, local coordinates for U, a submanifold of D'.

$$f_1^{\theta\mu}[f_1^{d+e+1} - f_2^e f_3^d] \in \Gamma(A, (\mathcal{I}^\mu)^{\theta\mu + d + e + 1})$$

since $f_1^\theta \in \Gamma(A, p_3^\theta \ldots p_n^\theta \mathcal{I}^{\theta\mu})$, $f_2^e \in \Gamma(A, p_2^e \mathcal{I}^{e\mu})$ and $f_3^d \in \Gamma(A, p_1^d \mathcal{I}^{d\mu})$. By Lemma 7.8 and Theorem 7.5, we can write $f_1^{\theta\mu}[f_1^{d+e+1} - f_2^e f_3^d]$ as a convergent power series in the f_j with no terms of order less than $\theta\mu + d + e + 1$. But $f_1^{\theta\mu}[f_1^{d+e+1} - f_2^e f_3^d] = [y^d x^e]^{\theta\mu + d + e + 1} k(x,y)$ with $k(0,0) = 0$ by (7.9) – (7.11). Thus the non-zero homogeneous terms in the power series expansion cannot include $f_1^{\theta\mu + d + e + 1}$. Thus as we derived (7.7), we derive

(7.12) $$v - u_2^e u_3^d - \Sigma a_\alpha v^{|\alpha| - \theta\mu - d - e} u^{\alpha'} = 0$$

and the summation contains only terms of at least second order. Thus v depends on quadratic and higher order terms.

Now for f_j, $j \geq 4$, $f_1 f_j$ has an image in $\Gamma(A, \mathcal{J}^{2\mu}/p_1 \mathcal{J}^{2\mu})$ which vanishes to second order at q. Hence, regarding second order vanishing in $\mathcal{J}^{2\mu}/p_1 \mathcal{J}^{2\mu}$ as the product of first order vanishing in $\mathcal{J}^{\mu}/p_1 \mathcal{J}^{\mu}$, we have by Lemma 7.3 that there exist constants a_{ke}, $k, e \geq 2$ such that $f_1 f_j - \Sigma\, a_{ke} f_k f_e \in \Gamma(A, p_1 \mathcal{J}^{2\mu})$. There are constants α and β such that $f_1 f_j - \Sigma\, a_{ke} f_k f_e - \alpha f_1 f_3 - \beta f_3^2$ vanishes at q to second order in $p_1 \mathcal{J}^{2\mu}/p_2 p_1 \mathcal{J}^{2\mu} \approx (p_1 \mathcal{J}^{\mu}/p_2 p_1 \mathcal{J}^{\mu}) \otimes (\mathcal{J}^{\mu}/p_2 \mathcal{J}^{\mu})$. By Lemma 7.3, there are constants $b_{\nu\tau}$ such that $\nu, \tau \geq 2$ and
$f_1 f_j - \Sigma\, a_{ke} f_k f_e - \alpha f_1 f_3 - \beta f_3^3 - \Sigma\, b_{\nu\tau} f_\nu f_\tau \in \Gamma(A, p_1 p_2 \mathcal{J}^{2\mu})$. The proof of Lemma 7.4 is easily modified to show that for sufficiently large μ, $\Gamma(A, p_1 \mathcal{J}^{\mu}/\mathcal{J}^{2\mu}) \otimes \Gamma(A, p_2 \mathcal{J}^{\mu}/\mathcal{J}^{2\mu}) \to \Gamma(A, p_1 p_2 \mathcal{J}^{2\mu}/\mathcal{J}^{3\mu})$ is surjective. So finally, there exist constants c_{mn}, $m, n \geq 2$ such that
$f_1 f_j - \Sigma\, a_{ke} f_k f_e - \alpha f_1 f_3 - \beta f_3^2 - \Sigma\, b_{\nu\tau} f_\nu f_\tau - \Sigma\, c_{mn} f_m f_n \in \Gamma(A, \mathcal{J}^{3\mu})$.
Then for appropriate $g(\zeta)$, having only third and higher order terms in its power series expansion, we get the following equation for W'.

$$u_j - \Sigma\, a_{ke} u_k u_e - \alpha u_3 - \beta u_3^2 - \Sigma\, b_{\nu\tau} u_\nu u_\tau - \Sigma\, c_{mn} u_m u_n - \frac{1}{v^2}\, g(v, u_2 v, \ldots, u_s v) = 0\,. \tag{7.13}$$

v depends on quadratic terms by (7.12). Hence (7.13) expresses u_j as αu_3 + (quadratic terms), as desired.

Proof of Theorem 7.1. Let $\phi : \mathcal{O}_p/m^\lambda \to \mathcal{O}_{\tilde{p}}/\tilde{m}^\lambda$ be the given isomorphism. λ will be estimated as we go along. Choose μ large enough so that all of the previous results of this section hold. Let $T = \Gamma(A, \mathcal{J}^{\mu})$. $m \supset T \supset m^{\mu\theta}$ by Corollary 7.6. Hence ϕ induces an isomorphism $\phi: T/T^{\tau} \to \phi(T)/(\phi(T))^{\tau}$ for $\tau \leq \lambda/\mu\theta$. Let $f_1, \ldots, f_s \in T$ have images in T/T^2 which form a basis of T/T^2. Let $\tilde{f}_i \in \mathcal{O}_{\tilde{p}}$ have $\phi(f_i + m^{\lambda})$ as its image in $\mathcal{O}_{\tilde{p}}/\tilde{m}^{\lambda}$. Appropriate polynomials in the f_i generate $m^{\mu\theta}/m^{2\mu\theta}$ so appropriate polynomials in the $\tilde{f}_i$ generate $m^{\mu\theta}/m^{2\mu\theta}$. Hence by Lemma 7.7, any element of $\tilde{m}^{\mu\theta}$ can be written as a convergent power series in the $\tilde{f}_i$.

We claim that there is a neighborhood N of $\tilde{p}$ such that $\tilde{F} = (\tilde{f}_1, \ldots, \tilde{f}_s) : N \to \tilde{W}$ is a proper map onto a subvariety $\tilde{W}$ of a polydisc in $\mathbf{C}^s$ and $\tilde{F}$ is biholomorphic off $\tilde{p}$. By neatly embedding p and $\tilde{p}$ we may assume that $\phi: \mathcal{O}_p/m^{\lambda} \to \mathcal{O}_{\tilde{p}}/\tilde{m}^{\lambda}$ is induced by an ambient isomorphism. Let $z_1, \ldots, z_t$ be the ambient coordinate functions. Then power series in the $\tilde{f}_i$ yield $z_1^{\mu\theta}, z_1^{\mu\theta-1}z_2, \ldots, z_1^{\mu\theta-1}z_t$, which are local coordinates for $z_1 \neq 0$. Thus the $\tilde{f}_i$ contain local coordinates for $z_1 \neq 0$, and similarly for $z_j \neq 0$, any j. Thus $\tilde{F}$ is biholomorphic off p for sufficiently small N. We show that $\tilde{F}$ is proper, as in the proof of Lemma 7.8, by composing $\tilde{F}$ with $\psi = (\psi_1, \ldots, \psi_t)$, $\psi_j(\tilde{f}_1, \ldots, \tilde{f}_s) = z_j^{\mu\theta}$. $\psi \circ \tilde{F}$ is a proper map into a polydisc for appropriate N so $\tilde{F}$ is a proper map into a polydisc for appropriate N. $\tilde{F}(N)$ is a subvariety $\tilde{W}$ by Theorem V.C.5 of G & R, the proper mapping theorem. We also have the subvariety W of Lemma 7.8. Think of W and $\tilde{W}$ as lying in the same polydisc in $\mathbf{C}^s$.

Perform a quadratic transformation in $\mathbf{C}^s$. $\pi|_{W'}: W' \to W$ is a resolution with A as exceptional set by construction. We also have $\pi|_{\tilde{W}'}: \tilde{W}' \to W'$. Let $\tilde{A} = \tilde{W}' \cap \pi^{-1}(0)$. As shown just before the proof of Lemma 7.9, A is the common set of zeroes for equations which depended only the the C-algebra structure of T/T^4 and $T/T^{\mu\theta+1}$. Using the isomorphism ϕ, we see that the same equations hold on $\tilde{A}$. Hence $\tilde{A} \subset A$ (but we do not yet have equality). At any point $q \in \tilde{A} \subset A$, we have s-2 equations, either (7.7) and (7.8) or (7.12) and (7.13), giving W' as a submanifold.

Using ϕ, we get corresponding equations for $\tilde{W}'$. Moreover, by Corollary 7.6, by increasing λ, we can increase indefinitely the orders to which the equations for W' and $\tilde{W}'$ agree. In particular we can preserve the linear terms. Hence near q, $\tilde{W}'$ is contained in a 2-dimensional manifold. $\tilde{p}$ is of pure dimension 2 so $\tilde{W}'$ is of pure dimension 2. Hence $\tilde{W}'$ is a submanifold defined by the equations corresponding to (7.7) and (7.8) or to (7.12) and (7.13) near q. $\tilde{A}$ is obtained by setting $v = 0$, as is A. But setting $v = 0$ gives identical equations for A and $\tilde{A}$ in (7.7) and (7.8) or (7.12) and (7.13). Hence $\tilde{A}$ is an open subset of A. But $\tilde{A}$ is compact and A is connected so that $A = \tilde{A}$. Thus $W'$and $\tilde{W}'$ are both submanifolds of D' and $W' \cap \pi^{-1}(0) = \tilde{W}' \cap \pi^{-1}(0)$.

Let $\mathcal{J}$ be the ideal sheaf of A in W' and $\tilde{\mathcal{J}}$ the ideal sheaf of $\tilde{A}$ in $\tilde{W}'$. We wish to determine for which ν the identity map on the ambient space D' induces an isomorphism $\psi : A(\mathcal{J}^{\nu}) \to \tilde{A}(\tilde{\mathcal{J}}^{\nu})$ between the non-reduced spaces $A(\mathcal{J}^{\nu})$ and $\tilde{A}(\tilde{\mathcal{J}}^{\nu})$ defined before Proposition 6.3. More precisely, let $\mathcal{G}$ and $\tilde{\mathcal{G}}$ be the ideal sheaves for W' and $\tilde{W}'$ in D'. Let $\mathcal{H}$ and $\tilde{\mathcal{H}}$ be the ideal sheaves generated by $\mathcal{G}$ and $\mathcal{J}^{\nu}$ and $\tilde{\mathcal{G}}$ and $\tilde{\mathcal{J}}^{\nu}$ respectively ($\mathcal{J}$ is defined modulo $\mathcal{G}$, so $\mathcal{H}$ is well defined.) It would suffice to determine for which ν, $\mathcal{H} = \tilde{\mathcal{H}}$, for then by passing to the quotients we would have the desired isomorphism ψ. We shall actually proceed in a slightly different manner.

$\mathcal{G}$ and $\tilde{\mathcal{G}}$ are generated near $A = \tilde{A}$ by (7.7) and (7.8) or (7.12) and (7.13). By requiring λ to be large, we can make the a_{α} in (7.7) or (7.12), whose left sides are among the generators of $\mathcal{G}$, and the $\tilde{a}_{\alpha}$ for $\tilde{\mathcal{G}}$ agree for $|\alpha| \leq \eta$, η large. Similarly g of (7.8) or (7.13) for $\mathcal{G}$ will agree with $\tilde{g}$ for $\tilde{\mathcal{G}}$ up to homogeneous terms in their power series of degree at most η, η arbitrarily large for arbitrarily large λ. Then the ideals generated by $\mathcal{G}$ and v^{ν} and by $\tilde{\mathcal{G}}$ and v^{ν} coincide for $\nu \leq \eta - 3\mu\theta$. On W', loc $v = A$ and on $\tilde{W}'$, loc $v = \tilde{A}$. Let (v) denote the ideal generated by v. Near q, say $q = A_1 \cap A_2$, $(v) = p_1^a p_2^b$ on W' for some a and b and $v = \tilde{p}_1^{\tilde{a}} \tilde{p}_2^{\tilde{b}}$ on W' for some $\tilde{a}$ and $\tilde{b}$. $\mathcal{O}/(\mathcal{G}, v^\nu) \approx \mathcal{O}/(\tilde{\mathcal{G}}, v^\nu)$, so we must have $a = \tilde{a}$ and $b = \tilde{b}$ as follows. On A_1, off the singular points, $p_1^{a-1} \not\equiv 0$ but $p_1^a \equiv 0$ and $\tilde{p}_1^{\tilde{a}-1} \not\equiv 0$ but $\tilde{p}_1^{\tilde{a}} \equiv 0$. Since we have a sheaf isomorphism on A, $a = \tilde{a}$. Similarly $b = \tilde{b}$. Since $a,b \geq 1$, we have $A(\mathcal{J}^\nu) \approx \tilde{A}(\mathcal{J}^\nu)$ for $\nu \leq \eta - 3\mu\theta$. By Theorem 6.20, for λ sufficiently large, A and $\tilde{A}$ have biholomorphic neighborhoods. Since p and $\tilde{p}$ are normal, they have biholomorphic neighborhoods by Theorem 3.13. ∎

Our last result says that hypersurface normal 2-dimensional singularities are algebraic.

COROLLARY 7.10. *Let* $f(x,y,z) = \Sigma\, a_{ijk} x^i y^j z^k$ *generate the ideal of a subvariety* $V \ni 0$. *Suppose that* 0 *is an isolated singularity of* V. *Then there exists an* N *such that for* $n \geq N$, $g(x,y,z) = \Sigma\, a_{ijk} x^i y^j z^k$, $i,j,k \leq n$, *generates the ideal of a subvariety* $\tilde{V}$ *such that* V *and* $\tilde{V}$ *have biholomorphic neighborhoods of* 0.

Proof: 0 is an isolated singularity of V, so that in some compact neighborhood $\overline{\Delta}$ of the origin in $\mathbb{C}^3$, we have that the power series of f converges in a neighborhood of $\overline{\Delta}$ and 0 is the only common zero of f, $\frac{\partial f}{\partial x}$, $\frac{\partial f}{\partial y}$, and $\frac{\partial f}{\partial z}$. Hence there exists an $\epsilon > 0$ such that on $\partial\Delta$, $\sup(|f|, |\frac{\partial f}{\partial x}|, |\frac{\partial f}{\partial y}|, |\frac{\partial f}{\partial z}|) > \varepsilon$. Require that N be sufficiently large so that on $\partial\Delta$, $\sup(|g|, |\frac{\partial g}{\partial x}|, |\frac{\partial g}{\partial y}|, |\frac{\partial g}{\partial z}|) > \frac{\varepsilon}{2}$. Thus in Δ, $V(g, \frac{\partial g}{\partial x}, \frac{\partial g}{\partial y}, \frac{\partial g}{\partial z})$ is discrete, so that 0 is an isolated singularity of $\tilde{V}$. (Of course, $\tilde{V}$ might have other singularities in Δ but these are also isolated.) g is square-free, for otherwise by Theorem II.E.19 of G & R, g would vanish to second order on some of the regular points of V(g) and there the partial derivatives of g would also vanish. Thus g generates $\mathscr{I}d(V(g))$ near 0. Letting m and $\tilde{m}$ be the maximal ideals in ${}_V\mathcal{O}_o$ and ${}_{\tilde{V}}\mathcal{O}_o$ respectively, we have a $\mathbb{C}$-algebra isomorphism ${}_V\mathcal{O}_o/\tilde{m}^N \approx {}_{\tilde{V}}\mathcal{O}_o/\tilde{m}^N$. Both V and $\tilde{V}$ have isolated and hence by Theorem 3.1 normal singularities at the origin. Therefore V and $\tilde{V}$ have biholomorphic neighborhoods of the origin for sufficiently large N by Theorem 7.1. ∎

BIBLIOGRAPHY

[A & G] Andreotti, A. and Grauert H., *Théorèmes de finitude pour la cohomologie des espaces complexes,* Bull. Soc. Math. France 90 (1962), 193-259.

[A] Artin, M., *Algebraic approximation of structures over complete local rings,* Inst. des Hautes Études Scientifiques, Publ. Math. n° 36 (1969) 23-58.

[B] Brieskorn, E., *Über die Auflösung gewisser Singularitäten von holomorphen Abbildungen,* Math. Ann. 166 (1966), 76-102.

[Gr1] Grauert, H., *Ein Theorem der analytischen Garbentheorie und die Modulräume komplexer Strukturen,* Inst. Hautes Études Scientifiques, Publ. Math. n° 5 (1960).

[Gr2] ______, *Über Modifikationen und exzeptionelle analytische Mengen,* Math. Ann. 146 (1962) 331-368.

[Gu] Gunning, R., *Lectures on Riemann Surfaces,* Princeton Univ. Press, Princeton, N. J. 1966.

[G&R] Gunning, R., and Rossi, H., *Analytic Functions of Several Complex Variables,* Prentice-Hall, Inc., Englewood Cliffs, N.J., 1965.

[Hℓ] Hille, E., *Analytic Function Theory,* Vol. II, Ginn and Co., Boston, 1962.

[Hr1] Hironaka, H., *Resolution of singularities of an algebraic variety over a field of characteristic zero: I, II,* Ann. Math., 79 (1964) 109-326.

[Hr2] ______, *a fundamental lemma on point modifications, Proceedings of the Conference on Complex Analysis, Minneapolis 1964,* Aeppli, et al. eds., Springer Verlag New York, Inc., 1965.

[H&R] Hironaka, H., and Rossi, H., *On the equivalence of imbeddings of exceptional complex spaces,* Math. Ann., 156 (1964) 313-333.

[Hz1] Hirzebruch, F., *Über vierdimensionale Riemannsche Flachen mehrdeutiger analytischer Funktionen von zwei komplexen Veranderlichen,* Math. Ann., 126 (1953) 1-22.

[Hz2] ———, *Differentiable Manifolds and Quadratic Forms,* Department of Mathematics, University of California, Berkeley, 1962.

[Ho] Hopf, H., *Schlichte Abbildungen und lokale Modifikationen 4-dimensionaler komplexer Mannigfaltigkeiten,* Comm. Math. Helv., 29 (1955) 132-156.

[K&N] Kelley, J., Namioka, N., et al., *Linear Topological Spaces,* Van Nostrand, Princeton, N. J., 1963.

[K] Knorr, K., *Über den Grauertschen Kohärenzsatz bei eigentlichen holomorphen Abbildungen, I, II,* Ann. Sci. École Norm. Sup. Pisa, 22 (1968) 729-768, 23 (1969) 1-74.

[M] Mumford, D., *The topology of normal singularities of an albegraic surface and a criterion for simplicity,* Inst. des Hautes Etudes Scientifiques, Publ. Math. n° 9 (1961) 5-22.

[N] Narasimhan, R., *Introduction to the Theory of Analytic Spaces,* Springer-Verlag, Berlin, 1966.

[O] Oka, K., *Sur les Fonctions Analytiques de Plusieurs Variables,* Iwanami Shoten, Tokyo, 1961, 127-157.

[Se] Serre, J-P., *Faisceaux algébriques coherents,* Ann. Math. 61 (1955) 197-278.

[Sp] Spanier, E., *Algebraic Topology,* McGraw-Hill, New York, 1966.

INDEX

QUEEN MARY
COLLEGE
LIBRARY